Measure Theoretic Laws
for lim sup Sets

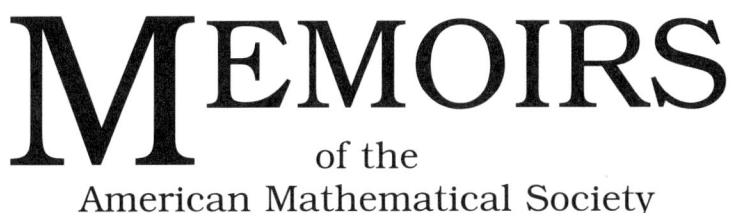

MEMOIRS
of the
American Mathematical Society

Number 846

Measure Theoretic Laws
for lim sup Sets

Victor Beresnevich
Detta Dickinson
Sanju Velani

January 2006 • Volume 179 • Number 846 (end of volume) • ISSN 0065-9266

American Mathematical Society
Providence, Rhode Island

2000 *Mathematics Subject Classification.*
Primary 11J83, 11J13, 11K60, 28A78, 28A80.

Library of Congress Cataloging-in-Publication Data

Beresnevich, Victor, 1971–
    Measure theoretic laws for lim sup sets / Victor Beresnevich, Detta Dickinson, Sanju Velani.
      p. cm. — (Memoirs of the American Mathematical Society, ISSN 0065-9266 ; no. 846)
    "Volume 179, number 846 (end of volume).
    Includes bibliographical references.
    ISBN 0-8218-3827-X (alk. paper)
    1. Diophantine approximation. 2. Probabilities. 3. Hausdorff measures. 4. Fractals.
I. Dickinson, Detta, 1968– II. Velani, Sanju, 1966– III. Title. IV. Series.

QA3.A57 no. 846
[QA242]
510s —dc22
[512.7′3]   2005053661

Memoirs of the American Mathematical Society

This journal is devoted entirely to research in pure and applied mathematics.

Subscription information. The 2006 subscription begins with volume 179 and consists of six mailings, each containing one or more numbers. Subscription prices for 2006 are US$624 list, US$499 institutional member. A late charge of 10% of the subscription price will be imposed on orders received from nonmembers after January 1 of the subscription year. Subscribers outside the United States and India must pay a postage surcharge of US$31; subscribers in India must pay a postage surcharge of US$43. Expedited delivery to destinations in North America US$35; elsewhere US$130. Each number may be ordered separately; *please specify number* when ordering an individual number. For prices and titles of recently released numbers, see the New Publications sections of the *Notices of the American Mathematical Society.*
    Back number information. For back issues see the *AMS Catalog of Publications.*
    Subscriptions and orders should be addressed to the American Mathematical Society, P. O. Box 845904, Boston, MA 02284-5904, USA. *All orders must be accompanied by payment.* Other correspondence should be addressed to 201 Charles Street, Providence, RI 02904-2294, USA.
    Copying and reprinting. Individual readers of this publication, and nonprofit libraries acting for them, are permitted to make fair use of the material, such as to copy a chapter for use in teaching or research. Permission is granted to quote brief passages from this publication in reviews, provided the customary acknowledgment of the source is given.
    Republication, systematic copying, or multiple reproduction of any material in this publication is permitted only under license from the American Mathematical Society. Requests for such permission should be addressed to the Acquisitions Department, American Mathematical Society, 201 Charles Street, Providence, Rhode Island 02904-2294, USA. Requests can also be made by e-mail to reprint-permission@ams.org.

Memoirs of the American Mathematical Society is published bimonthly (each volume consisting usually of more than one number) by the American Mathematical Society at 201 Charles Street, Providence, RI 02904-2294, USA. Periodicals postage paid at Providence, RI. Postmaster: Send address changes to Memoirs, American Mathematical Society, 201 Charles Street, Providence, RI 02904-2294, USA.

10 9 8 7 6 5 4 3 2 1   11 10 09 08 07 06

Dedicated to Bridget, Ayesha and Iona

Contents

Abstract

Given a compact metric space (Ω, d) equipped with a non-atomic, probability measure m and a positive decreasing function ψ, we consider a natural class of lim sup subsets $\Lambda(\psi)$ of Ω. The classical lim sup set $W(\psi)$ of 'ψ–approximable' numbers in the theory of metric Diophantine approximation fall within this class. We establish sufficient conditions (which are also necessary under some natural assumptions) for the m–measure of $\Lambda(\psi)$ to be either positive or full in Ω and for the Hausdorff f-measure to be infinite. The classical theorems of Khintchine-Groshev and Jarník concerning $W(\psi)$ fall into our general framework. The main results provide a unifying treatment of numerous problems in metric Diophantine approximation including those for real, complex and p-adic fields associated with both independent and dependent quantities. Applications also include those to Kleinian groups and rational maps. Compared to previous works our framework allows us to successfully remove many unnecessary conditions and strengthen fundamental results such as Jarník's theorem and the Baker-Schmidt theorem. In particular, the strengthening of Jarník's theorem opens up the Duffin-Schaeffer conjecture for Hausdorff measures.

Mathematics Subject Classification: 11J83; 11J13, 11K60, 28A78, 28A80

Keywords: Metric Diophantine approximation, Hausdorff measure and dimension, limsup sets, Khintchine and Jarník theorems, zero–one laws,

Received by the editor September 14 2004

V Beresnevich's work was partially supported by INTAS project 00-429

S Velani is a Royal Society University Research Fellow

1. Introduction

1.1. Background: the basic example.

To set the scene for the abstract framework considered in this article we introduce a basic $\limsup$ set whose study has played a central role in the development of the classical theory of metric Diophantine approximation. Given a real, positive decreasing function $\psi : \mathbb{R}^+ \to \mathbb{R}^+$, let

$$W(\psi) := \{x \in [0,1] : |x - p/q| < \psi(q) \text{ for i.m. rationals } p/q \ (q > 0)\},$$

where 'i.m.' means 'infinitely many'. This is the classical set of ψ–well approximable numbers in the theory of one dimensional Diophantine approximation. The fact that we have restricted our attention to the unit interval rather than the real line is purely for convenience. It is natural to refer to the function ψ as the *approximating function*. It governs the 'rate' at which points in the unit interval must be approximated by rationals in order to lie in $W(\psi)$. It is not difficult to see that $W(\psi)$ is a $\limsup$ set. For $n \in \mathbb{N}$, let

$$W(\psi, n) := \bigcup_{k^{n-1} < q \leq k^n} \bigcup_{0 \leq p \leq q} B(p/q, \psi(q)) \cap [0,1]$$

where $k > 1$ is fixed and $B(c,r)$ is the open interval centred at c of radius r. The set $W(\psi)$ consists precisely of points in the unit interval that lie in infinitely many $W(\psi, n)$; that is

$$W(\psi) = \limsup_{n \to \infty} W(\psi, n) := \bigcap_{m=1}^{\infty} \bigcup_{n=m}^{\infty} W(\psi, n) \ .$$

Investigating the measure theoretic properties of the set $W(\psi)$ underpins the classical theory of metric Diophantine approximation. We begin by considering the 'size' of $W(\psi)$ expressed in terms of the ambient measure m; i.e. one–dimensional Lebesgue measure. On exploiting the $\limsup$ nature of $W(\psi)$, a straightforward application of the convergence part of the Borel–Cantelli lemma from probability theory yields that

$$m(W(\psi)) = 0 \quad \text{if} \quad \sum_{n=1}^{\infty} k^{2n} \psi(k^n) \ < \ \infty \ .$$

Notice that since ψ is monotonic, the convergence/divergence property of the above sum is equivalent to that of $\sum_{r=1}^{\infty} r \, \psi(r)$.

A natural problem now arises. Under what conditions is $m(W(\psi)) > 0$? The following fundamental result provides a beautiful and simple criteria for the 'size' of the set $W(\psi)$ expressed in terms of Lebesgue measure.

KHINTCHINE'S THEOREM (1924). *If $\psi(r)$ is decreasing then*

$$m(W(\psi)) = \begin{cases} 0 & \text{if} \quad \sum_{r=1}^{\infty} r\,\psi(r) < \infty\,, \\ 1 & \text{if} \quad \sum_{r=1}^{\infty} r\,\psi(r) = \infty\,. \end{cases}$$

Thus, in the divergence case, which constitutes the main substance of Khintchine's theorem, not only do we have positive Lebesgue measure but full Lebesgue measure. In fact, this turns out to be the case for all the examples considered in this paper. Usually, there is a standard argument which allows one to deduce full measure from positive measure – such as the invariance of the lim sup set or some related set, under an ergodic transformation. In any case, we shall prove a general result which directly implies the above full measure statement. It is worth mentioning that in Khintchine's original statement the stronger hypothesis that $r^2\psi(r)$ is decreasing was assumed.

Returning to the convergence case, we cannot obtain any further information regarding the 'size' of $W(\psi)$ in terms of Lebesgue measure — it is always zero. Intuitively, the 'size' of $W(\psi)$ should decrease as the rate of approximation governed by the function ψ increases. In short, we require a more delicate notion of 'size' than simply Lebesgue measure. The appropriate notion of 'size' best suited for describing the finer measure theoretic structures of $W(\psi)$ is that of generalized Hausdorff measures. The Hausdorff f–measure $\mathcal{H}^f$ with respect to a dimension function f is a natural generalization of Lebesgue measure. So as not to interrupt the flow of this background/motivation exposition we referee the reader to §7 for the standard definition of $\mathcal{H}^f$ and further comments regarding Hausdorff measures and dimension.

Again on exploiting the lim sup nature of $W(\psi)$, a straightforward covering argument provides a simple convergence condition under which $\mathcal{H}^f(W(\psi)) = 0$. Thus, in view of the development of the Lebesgue theory it is natural to ask for conditions under which $\mathcal{H}^f(W(\psi))$ is strictly positive.

The following fundamental result provides a beautiful and simple criteria for the 'size' of the set $W(\psi)$ expressed in terms of Hausdorff measures.

JARNÍK'S THEOREM (1931). *Let f be a dimension function such that $r^{-1} f(r) \to \infty$ as $r \to 0$ and $r^{-1} f(r)$ is decreasing. Let ψ be a real, positive decreasing function. Then*

$$\mathcal{H}^f\left(W(\psi)\right) = \begin{cases} 0 & \text{if} \quad \sum_{r=1}^{\infty} \ r f\left(\psi(r)\right) < \infty\,, \\ \infty & \text{if} \quad \sum_{r=1}^{\infty} \ r f\left(\psi(r)\right) = \infty\,. \end{cases}$$

Clearly the above theorem can be regarded as the Hausdorff measure version of Khintchine's theorem. As with the latter, the divergence part constitutes the main substance. Notice, that the case when $\mathcal{H}^f$ is comparable to one–dimensional Lebesgue measure m (i.e. $f(r) = r$) is excluded by the condition $r^{-1} f(r) \to \infty$ as $r \to 0$. Analogous to Khintchine's original statement, in Jarník's original statement the additional hypotheses that $r^2 \psi(r)$ is decreasing, $r^2 \psi(r) \to 0$ as $r \to \infty$ and that $r^2 f(\psi(r))$ is decreasing were assumed. Thus, even in the simple case when $f(r) = r^s$ $(s \geq 0)$ and the approximating function is given by $\psi(r) = r^{-\tau} \log r$ $(\tau > 2)$, Jarník's original statement gives no information regarding the s–dimensional Hausdorff measure of $W(\psi)$ at the critical exponent $s = 2/\tau$ – see below. That this is the case is due to the fact that $r^2 f(\psi(r))$ is not decreasing. However, as we shall see these additional hypotheses are unnecessary. More to the point, Jarník's theorem as stated above is the precise Hausdorff measure version of Khintchine's theorem. Of course, as with Khintchine's theorem the question of removing the monotonicity condition on the approximating function ψ now arises. That is to say, it now makes perfect sense to consider a generalized Duffin-Schaeffer conjecture for Hausdorff measures – for a detailed account regarding the original Duffin-Schaeffer conjecture see [22, 39]. Briefly, let $\psi(n)$ be a sequence of non-negative real numbers and consider the set $\widetilde{W}(\psi)$ of $x \in [0,1]$ for which there exist infinitely many rationals p/q $(q \geq 1)$ such that

$$|x - p/q| \ < \ \psi(q) \qquad \text{with} \qquad (p, q) = 1\,.$$

The Duffin-Schaeffer conjecture for Hausdorff measures: Let f be a dimension function such that $r^{-1} f(r) \to \infty$ as $r \to 0$ and $r^{-1} f(r)$ is decreasing. Let ϕ denote the Euler function. Then

$$\mathcal{H}^f\left(\widetilde{W}(\psi)\right) = \infty \quad \text{if} \quad \sum_{n=1}^{\infty} f\left(\psi(n)\right) \phi(n) \ = \infty\,.$$

It is easy to show that $\mathcal{H}^f(\widetilde{W}(\psi)) = 0$ if the above sum converges. The higher dimensional Duffin-Schaeffer conjecture corresponding to simultaneous approximation is known to be true [**37**]. It is plausible that the ideas in [**37**] together with those in this paper are sufficient to prove the higher dimensional version of the above conjecture. The first and last authors have shown that this is indeed the case [**9**].

Returning to Jarník's theorem, note that in the case when $\mathcal{H}^f$ is the standard s–dimensional Hausdorff measure $\mathcal{H}^s$ (i.e. $f(r) = r^s$), it follows from the definition of Hausdorff dimension (see §7) that

$$\dim W(\psi) = \inf\{s : \textstyle\sum_{r=1}^{\infty} r\,\psi(r)^s < \infty\} .$$

Previously, Jarník (1929) and independently Besicovitch (1934) had determined the Hausdorff dimension of the set $W(r \mapsto r^{-\tau})$, usually denoted by $W(\tau)$, of τ–well approximable numbers. They proved that for $\tau > 2$, $\dim W(\tau) = 2/\tau$. Thus, as the 'rate' of approximation increases (i.e. as τ increases) the 'size' of the set $W(\tau)$ expressed in terms of Hausdorff dimension decreases. As discussed earlier, this is in precise keeping with one's intuition. Obviously, the dimension result implies that

$$\mathcal{H}^s\left(W(\tau)\right) = \begin{cases} 0 & \text{if } s > 2/\tau \\ \infty & \text{if } s < 2/\tau \end{cases} ,$$

but gives no information regarding the s–dimensional Hausdorff measure of $W(\tau)$ at the critical value $s = \dim W(\tau)$. Clearly, Jarník's zero–infinity law implies the dimension result and that for $\tau > 2$

$$\mathcal{H}^{2/\tau}(W(\tau)) = \infty .$$

Furthermore, the 'zero–infinity' law allows us to discriminate between sets with the same dimension and even the same s–dimensional Hausdorff measure. For example, with $\tau \geq 2$ and $0 < \epsilon_1 < \epsilon_2$ consider the approximating functions

$$\psi_{\epsilon_i}(r) := r^{-\tau}\,(\log r)^{-\frac{\tau}{2}(1+\epsilon_i)} \qquad (i = 1, 2) .$$

It is easily verified that for any $\epsilon_i > 0$,

$$m(W(\psi_{\epsilon_i})) = 0 , \quad \dim W(\psi_{\epsilon_i}) = 2/\tau \quad \text{and} \quad \mathcal{H}^{2/\tau}(W(\psi_{\epsilon_i})) = 0 .$$

However, consider the dimension function f given by $f(r) = r^{2/\tau}(\log r^{-1/\tau})^{\epsilon_1}$. Then $\sum_{r=1}^{\infty} r\,f\,(\psi_{\epsilon_i}(r)) \asymp \sum_{r=1}^{\infty} (r\,(\log r)^{1+\epsilon_i-\epsilon_1})^{-1}$, where as usual the symbol $\asymp$ denotes comparability (the quotient of the associated quantities is bounded from above and below by positive, finite constants). Hence, Jarník's zero–infinity law implies that

$$\mathcal{H}^f\left(W(\psi_{\epsilon_1})\right) = \infty \qquad \text{whilst} \qquad \mathcal{H}^f\left(W(\psi_{\epsilon_2})\right) = 0 .$$

Thus the Hausdorff measure $\mathcal{H}^f$ does make a distinction between the 'sizes' of the sets under consideration; unlike s–dimensional Hausdorff measure.

Within this classical setup, it is apparent that Khintchine's theorem together with Jarník's zero–infinity law provide a complete measure theoretic description of $W(\psi)$. In short, our central aim is to establish analogues of the divergence parts of these classical results within a general framework. Recall, that the divergence parts constitute the main substance of the classical statements.

1.2. The general setup and fundamental problems. Let (Ω, d) be a compact metric space equipped with a non-atomic, probability measure m. Let $\mathcal{R} = \{R_\alpha \subset \Omega : \alpha \in J\}$ be a family of subsets R_α of Ω indexed by an infinite, countable set J. The sets R_α will be referred to as *resonant sets* for reasons which will become apparent later. Next, let $\beta : J \to \mathbb{R}^+ : \alpha \mapsto \beta_\alpha$ be a positive function on J. Thus, the function β attaches a 'weight' β_α to the resonant set R_α. To avoid pathological situations within our framework, we shall assume that the number of α in J with β_α bounded above is always finite. For a set $A \subset \Omega$, let

$$\Delta(A, \delta) := \{x \in \Omega : \text{dist}\,(x, A) < \delta\}$$

where $\text{dist}\,(x, A) := \inf\{d(x, a) : a \in A\}$. Thus, $\Delta(A, \delta)$ is simply the δ-neighborhood of A. Given a decreasing function $\psi : \mathbb{R}^+ \to \mathbb{R}^+$ let

$$\Lambda(\psi) = \{x \in \Omega : x \in \Delta(R_\alpha, \psi(\beta_\alpha)) \text{ for infinitely many } \alpha \in J\}\,.$$

The set $\Lambda(\psi)$ is a 'lim sup' set; it consists of points in Ω which lie in infinitely many of the 'thickenings' $\Delta(R_\alpha, \psi(\beta_\alpha))$. Clearly, even in this abstract setup it is natural to refer to the function ψ as the *approximating function*. It governs the 'rate' at which points in Ω must be approximated by resonant sets in order to lie in $\Lambda(\psi)$. Notice, that in the case the resonant sets are points, the thickenings $\Delta(R_\alpha, \psi(\beta_\alpha))$ are simply balls $B(R_\alpha, \psi(\beta_\alpha))$ centred at resonant points.

Before continuing our discussion, we rewrite $\Lambda(\psi)$ in a fashion which brings its 'lim sup' nature to the forefront. For $n \in \mathbb{N}$, let

$$\Delta(\psi, n) := \bigcup_{\alpha \in J\,:\,k^{n-1} < \beta_\alpha \leq k^n} \Delta(R_\alpha, \psi(\beta_\alpha)) \quad \text{where } k > 1 \text{ is fixed.}$$

By assumption the number of α in J with $k^{n-1} < \beta_\alpha \leq k^n$ is finite regardless of the value of k. Thus, $\Lambda(\psi)$ is precisely the set of points in Ω which lie in infinitely many

$\Delta(\psi, n)$; that is

$$\Lambda(\psi) = \limsup_{n \to \infty} \Delta(\psi, n) := \bigcap_{m=1}^{\infty} \bigcup_{n=m}^{\infty} \Delta(\psi, n) \ .$$

The main line of our investigation is motivated by the following pair of fundamental problems regarding the measure theoretic structure of $\Lambda(\psi)$. In turn the
fundamental problems are motivated by the classical theory described in the previous section. It is reasonably straightforward to determine conditions under which
$m(\Lambda(\psi)) = 0$. In fact, this is implied by the convergence part of the Borel–Cantelli
lemma from probability theory whenever

$$(1) \qquad\qquad \sum_{n=1}^{\infty} m(\Delta(\psi, n)) < \infty \ .$$

In view of this it is natural to consider:

PROBLEM 1. *Under what conditions is* $m(\Lambda(\psi))$ *strictly positive ?*

Under a 'global ubiquity' hypothesis and a divergent sum condition, together with
mild conditions on the measure, our first theorem provides a complete solution to this
problem. Moreover, if we replace the 'global ubiquity' hypothesis by a 'local ubiquity'
hypothesis then $\Lambda(\psi)$ has full m–measure and this statement can be viewed as the
analogue of Khintchine's theorem or more generally as the analogue of the classical
linear forms theorem of Khintchine–Groshev.

Reiterating the above measure zero statement, if the approximating function ψ
decreases sufficiently quickly so that (1) is satisfied, the corresponding $\limsup$ set
$\Lambda(\psi)$ is of zero m–measure. As with the classical setup of §1.1, in this case we cannot
obtain any further information regarding the 'size' of $\Lambda(\psi)$ in terms of the ambient
measure m — it is always zero. In short, we require a more delicate notion of 'size'
than simply the given m-measure. In keeping with the classical development, we
investigate the 'size' of $\Lambda(\psi)$ with respect to the Hausdorff measures $\mathcal{H}^f$ where f
is a dimension function. Again, provided a certain 'f-volume' sum converges, it is
reasonably simple to determine conditions under which $\mathcal{H}^f(\Lambda(\psi)) = 0$. Naturally, we
consider:

PROBLEM 2. *Under what conditions is* $\mathcal{H}^f(\Lambda(\psi))$ *strictly positive ?*

This problem turns out to be far more subtle than the previous one regarding m-measure. To make any substantial progress, we impose the condition that the m-measure of any ball centred at a point in Ω is comparable to some fixed power of its radius. Then, under a 'local ubiquity' hypothesis and an 'f-volume' divergent sum condition, together with mild conditions on the dimension function, our second theorem shows that $\mathcal{H}^f(\Lambda(\psi)) = \infty$. Thus, $\mathcal{H}^f(\Lambda(\psi))$ satisfies an elegant 'zero–infinity' law whenever the convergence of the 'f-volume' sum implies $\mathcal{H}^f(\Lambda(\psi)) = 0$ as is often the case. In particular, this latter statement is true for the standard s-dimensional Hausdorff measure $\mathcal{H}^s$. Thus, in the language of geometric measure theory the sets $\Lambda(\psi)$ are not s-sets. Furthermore, from such zero–infinity laws it is easy to deduce the Hausdorff dimension of $\Lambda(\psi)$.

Examples of lim sup sets which fall into the above abstract framework include the classical sets of well approximable numbers/vectors in the theory of Diophantine approximation as well as the 'shrinking target' sets associated with the phase space of a given dynamical system.

In order to illustrate and clarify the above setup and our line of investigation, we return to the basic lim sup set of §1.1. The classical set $W(\psi)$ of ψ–well approximable numbers in the theory of one dimensional Diophantine approximation can clearly be expressed in the form $\Lambda(\psi)$ with

$$\Omega := [0,1] , \quad J := \{(p,q) \in \mathbb{N} \times \mathbb{N} : 0 \le p \le q\} , \quad \alpha := (p,q) \in J ,$$

$$\beta_\alpha := q , \quad R_\alpha := p/q \quad \text{and} \quad \Delta(R_\alpha, \psi(\beta_\alpha)) := B(p/q, \psi(q)) .$$

The metric d is of course the standard Euclidean metric; $d(x,y) := |x - y|$. Thus in this basic example, the resonant sets R_α are simply rational points p/q. Furthermore,

$$\Delta(\psi, n) := \bigcup_{k^{n-1} < q \le k^n} \bigcup_{p=0}^{q} B(p/q, \psi(q))$$

and $W(\psi) = \limsup \Delta(\psi, n)$ as $n \to \infty$.

For this basic example, the solution to our first fundamental problem is given by Khintchine's theorem and the solution to the second by Jarník's theorem. Together, these theorems provide a complete measure theoretic description of $W(\psi)$. In the case of the general framework, analogues of these results should be regarded as the ultimate pair of results describing the metric structure of the lim sup sets $\Lambda(\psi)$. Alternatively, they provide extremely satisfactory solutions to the fundamental problems. Analogues of the convergence parts of the classical results usually follow by adapting the 'natural cover'

$$\{\Delta(\psi, n) : n = m, m + 1, \cdots\} \qquad\qquad (m \in \mathbb{N})$$

of $\Lambda(\psi)$. Our key aim is to establish analogues of the divergence parts of the classical results for $\Lambda(\psi)$.

2. Ubiquity and conditions on the general setup

In order to make any reasonable progress with the fundamental problems we impose various conditions on the metric measure space (Ω, d, m). Moreover, we require the notion of a 'global' and 'local' ubiquitous system which will underpin our line of investigation. The general setup is independent of the approximating function ψ.

Throughout, a ball centred at a point x and radius r is defined to be the set $\{y \in \Omega : d(x, y) < r\}$ or $\{y \in \Omega : d(x, y) \leq r\}$ depending on whether it is open or closed. In general, we do not specify whether a certain ball is open or close since it will be irrelevant. Notice, that by definition any ball is automatically a subset of Ω.

2.1. Upper and lower sequences and the sets $J_l^u(n)$. Let $l := \{l_n\}$ and $u := \{u_n\}$ be positive increasing sequences such that

$$l_n < u_n \qquad\qquad \text{and} \qquad\qquad \lim_{n \to \infty} l_n = \infty .$$

Thus, $\lim_{n \to \infty} u_n = \infty$. Now, define

$$\Delta_l^u(\psi, n) := \bigcup_{\alpha \in J_l^u(n)} \Delta(R_\alpha, \psi(\beta_\alpha))$$

where

$$J_l^u(n) := \{\alpha \in J : l_n < \beta_\alpha \leq u_n\}.$$

By assumption the cardinality of $J_l^u(n)$ is finite regardless of l and u. In view of this and the fact that $l_n \to \infty$ as $n \to \infty$, it follows that

$$\Lambda(\psi) = \limsup_{n \to \infty} \Delta_l^u(\psi, n) := \bigcap_{m=1}^{\infty} \bigcup_{n=m}^{\infty} \Delta_l^u(\psi, n) .$$

This statement is irrespective of the choice of the sequences l and u. Note that without additional assumptions, the fact that $l_n \to \infty$ as $n \to \infty$ is crucial. For obvious reasons, the sequence l will be referred to as *the lower sequence* and u as *the upper sequence*.

2.2. The conditions on the measure and the space. The two central conditions on the measure m are as follows and will always be assumed throughout. Firstly, the m–measure of any ball centred at a point of the space Ω is strictly positive; i.e. $m(B(x, r)) > 0$ for x in Ω and $r > 0$. Secondly, the measure m is *doubling*. That is to say that there exists a constant $C \geq 1$ such that for any x in Ω

$$m(B(x, 2r)) \leq C\, m(B(x, r)) \ .$$

The doubling condition allows us to blow up a given ball by a constant factor without drastically affecting its measure. Also note that it implies that $m(B(x, tr)) \leq C(t)\, m(B(x, r))$ for any $t > 1$ with x in Ω. The metric measure space (Ω, d, m) is also said to be doubling if m is doubling [**23**].

Regarding 'Problem 1', we shall impose the following reasonably mild conditions on the measure. Essentially, it asserts that balls of the same radius centred at points on resonant sets R_α with $\alpha \in J_l^u(n) := \{\alpha \in J : l_n < \beta_\alpha \leq u_n\}$ have roughly the same measure for some choice of l and u.

(M1)   For $c \in R_\alpha$, $c' \in R_{\alpha'}$ with $\alpha, \alpha' \in J_l^u(n)$ and $r < r_o$

(2)
$$a \ \leq \ \frac{m(B(c, r))}{m(B(c', r))} \ \leq \ b \ ,$$

where the constants $a, b > 0$ are independent of n and the balls under consideration, but may depend on the l and u.

Regarding 'Problem 2', more is required. Namely, that the measure of a ball centred at a point in Ω is comparable to some fixed power of its radius.

(M2)   There exist positive constants δ and r_o such that for any $x \in \Omega$ and $r \leq r_o$,

(3)
$$a\, r^\delta \ \leq \ m(B(x, r)) \ \leq \ b\, r^\delta \ .$$

The constants a and b are independent of the ball and without loss of generality we assume that $0 < a < 1 < b$. Notice that if m satisfies condition (M2), then (M1) is trivially satisfied for any choice of l and u, as are the central conditions. Also, (M2) implies that $\dim \Omega = \delta$ – see §7 for the details.

2.3. The intersection conditions. In the case that the resonant sets are not points, we will require measure theoretic control on the intersection of certain balls centred at points on resonant sets with certain 'thickenings' of the resonant sets. The radii of the balls and the thickenings of the resonant sets are governed by a positive function ρ which is intimately tied up with the notion of ubiquity – see below.

THE INTERSECTION CONDITIONS There exists a constant γ with $0 \leq \gamma \leq \dim \Omega$, such that for any $\alpha \in J$ with $\beta_\alpha \leq u_n$, $c \in R_\alpha$ and $0 < \lambda \leq \rho(u_n)$ the following are satisfied for n sufficiently large:

$$(i) \quad m(B(c, \tfrac{1}{2}\rho(u_n)) \cap \Delta(R_\alpha, \lambda)) \geq c_1 \, m(B(c, \lambda)) \left(\frac{\rho(u_n)}{\lambda} \right)^\gamma$$

$$(ii) \quad m(B \cap B(c, 3\rho(u_n)) \cap \Delta(R_\alpha, 3\lambda)) \leq c_2 \, m(B(c, \lambda)) \left(\frac{r(B)}{\lambda} \right)^\gamma$$

where B is an arbitrary ball centred on a resonant set with radius $r(B) \leq 3\,\rho(u_n)$. The constants c_1 and c_2 are positive and absolute. Without loss of generality we assume that $0 < c_1 < 1 < c_2$.

When the resonant sets are points so that $\Delta(R_\alpha, \lambda) := B(c, \lambda)$, the above conditions are trivially satisfied with $\gamma = 0$. In applications, when the resonant sets are not points they are usually subsets of smooth manifolds or simply planes, all of the same dimension. In such cases, the intersection conditions hold with $\gamma = \dim R_\alpha$. In particular, it is readily verified that when (Ω, d) is a subspace of $\mathbb{R}^n$ and the resonant sets are γ–dimensional affine subspaces of Ω then the intersection conditions are inevitably satisfied. In view of this we refer to γ as the *common dimension* of the resonant sets in $\mathcal{R}$.

2.4. The ubiquitous systems. The following 'systems' contain the key measure theoretic structure necessary for our attack on the fundamental problems. Recall that $\mathcal{R}$ denotes the family of resonant sets R_α and that the function β attaches a 'weight' β_α to each resonant set $R_\alpha \in \mathcal{R}$.

Let $\rho : \mathbb{R}^+ \to \mathbb{R}^+$ be a function with $\rho(r) \to 0$ as $r \to \infty$ and let

$$\Delta_l^u(\rho, n) := \bigcup_{\alpha \in J_l^u(n)} \Delta(R_\alpha, \rho(u_n)) \ .$$

Definition (Local m–ubiquity) Let $B = B(x, r)$ be an arbitrary ball with centre x in Ω and radius $r \le r_0$. Suppose there exists a function ρ, sequences l and u and an absolute constant $\kappa > 0$ such that

(4) $$m\left(B \cap \Delta_l^u(\rho, n)\right) \ge \kappa \, m(B) \qquad \text{for } n \ge n_o(B).$$

Furthermore, suppose the intersection conditions are satisfied. Then the pair $(\mathcal{R}, \beta)$ is said to be a *local m-ubiquitous system relative to* (ρ, l, u).

Definition (Global m–ubiquity) Suppose there exists a function ρ, sequences l and u and an absolute constant $\kappa > 0$ such that for $n \ge n_o$, (4) is satisfied for $B := \Omega$. Furthermore, suppose the intersection conditions are satisfied. Then the pair $(\mathcal{R}, \beta)$ is said to be a *global m-ubiquitous system relative to* (ρ, l, u).

The function ρ, in either form of ubiquity will be referred to as the *ubiquitous function*. It is clear that for global ubiquity (4) reduces to $m\left(\Delta_l^u(\rho, n)\right) \ge \kappa$. Essentially, in the 'global' case all that is required is that the set $\Delta_l^u(\rho, n)$ approximates the underlying space Ω in terms of the measure m. In the 'local' case, this approximating property is required to hold locally on balls centred at points in Ω.

Clearly, local m–ubiquity implies global m–ubiquity. Simply take a ball B centred at a point of Ω with radius $\le r_o$, then for n sufficiently large $m\left(\Delta_l^u(\rho, n)\right) \ge m\left(B \cap \Delta_l^u(\rho, n)\right) \ge \kappa \, m(B) := \kappa_1 > 0$. In other words local ubiquity with a constant κ implies global ubiquity with some constant κ_1 where $0 < \kappa_1 \le \kappa$. In general the converse is not true. However, if

$$m\left(\Delta_l^u(\rho, n)\right) \ \to \ 1 \ = \ m(\Omega) \qquad \text{as} \qquad n \to \infty \ ,$$

then it is easy to show that global m–ubiquity implies local m–ubiquity. To see this, let B be any ball and assume without loss of generality that $m(B) = \epsilon > 0$. Then, for n sufficiently large $m\left(\Delta_l^u(\rho, n)\right) > m(\Omega) - \epsilon/2$. Hence, $m\left(B \cap \Delta_l^u(\rho, n)\right) \ge \epsilon/2$ as required. This rather simple observation can be extremely useful when trying to establish that a given system is locally m-ubiquitous.

In attempting to establish the measure theoretic inequality in either form of ubiquity, the presence of the lower sequence l is irrelevant. To see this, suppose we are able to show that for $n \geq n_o(B)$,

$$m\left(B \cap \bigcup_{\alpha \in J : \beta_\alpha \leq u_n} \Delta(R_\alpha, \rho(u_n))\right) \geq \kappa\, m(B) \ .$$

Since $\rho(r) \to 0$ as $r \to \infty$, for any $t \in \mathbb{N}$ there exists an integer n_t such that for $n \geq n_t$

$$m\left(B \cap \bigcup_{\alpha \in J : \beta_\alpha \leq t} \Delta(R_\alpha, \rho(u_n))\right) < \tfrac{1}{2} \kappa\, m(B) \ .$$

Without loss of generality we can assume that $n_{t+1} \geq n_t + 1$. Now consider the lower sequence l given by $l_n := t$ for $n \in [n_t, n_{t+1})$. Clearly, l is an increasing sequence with $l_n \to \infty$ as $n \to \infty$. Moreover, for n sufficiently large we have that $m\left(B \cap \Delta^u_l(\rho, n)\right) \geq \tfrac{1}{2}\kappa\, m(B)$; i.e. the pair $(\mathcal{R}, \beta)$ is a local m-ubiquitous system relative to (ρ, l, u).

The above discussion indicates that the lower sequence l is irrelevant within the ubiquity framework. Regarding the upper sequence, notice that any subsequence s of u will also do; i.e. the measure theoretic inequalities are satisfied for $\Delta^s_l(\rho, n)$. To see that this is the case, simply observe that for each $t \in \mathbb{N}$ we have that $s_t = u_n$ for some $n \geq t$. Then, $l_t \leq l_n$ and so $J^s_l(t) \supseteq J^u_l(n)$. Thus $\Delta^s_l(\rho, t) \supseteq \Delta^u_l(\rho, n)$, and it follows that $m\left(B \cap \Delta^s_l(\rho, t)\right) \geq m\left(B \cap \Delta^u_l(\rho, n)\right) \geq \kappa\, m(B)$; i.e. the pair $(\mathcal{R}, \beta)$ is a local m-ubiquitous system relative to (ρ, l, s).

In practice, the global or local m–ubiquity of a system can be established using standard arguments concerning the distribution of the resonant sets in Ω, from which the function ρ arises naturally. To illustrate this, we return to our basic example.

The basic example again: For the set $W(\psi)$ of ψ–well approximable numbers the resonant sets are simply rational points. Thus the intersection conditions are automatically satisfied with $\gamma = 0$. Of course, the measure m is one–dimensional Lebesgue measure and satisfies the measure condition (M2) with $\delta = 1$.

LEMMA 1. *There is a constant $k > 1$ such that the pair $(\mathcal{R}, \beta)$ is a local m-ubiquitous system relative to (ρ, l, u) where $l_{n+1} = u_n := k^n$ and $\rho : r \mapsto \text{constant} \times r^{-2}$.*

Proof. Let $I = [a, b] \subset [0, 1]$. By Dirichlet's theorem, for any $x \in I$ there are coprime integers p, q with $1 \leq q \leq k^n$ satisfying $|x - p/q| < (qk^n)^{-1}$. Clearly,

$aq - 1 \leq p \leq bq + 1$. Thus, for a fixed q there are at most $m(I)q + 3$ possible values of p. Trivially, for n large

$$m\left(I \cap \bigcup_{q \leq k^{n-1}} \bigcup_{p} B\left(\tfrac{p}{q}, \tfrac{1}{qk^n}\right)\right) \leq 2 \sum_{q \leq k^{n-1}} \tfrac{1}{qk^n} \left(m(I)q + 3\right) \leq \tfrac{3}{k} \, m(I).$$

It follows that for $k \geq 6$,

$$m\left(I \cap \bigcup_{k^{n-1} < q \leq k^n} \bigcup_{p} B\left(\tfrac{p}{q}, \tfrac{k}{k^{2n}}\right)\right) \geq m(I) - \tfrac{3}{k} \, m(I) \geq \tfrac{1}{2} \, m(I).$$

$$\#$$

It will be evident from our 'ubiquity' theorems, that Lemma 1 is sufficient for directly establishing the divergence part of both Khintchine's theorem and Jarník's zero–infinity law – see §6.

2.5. A remark on related systems. In the case that Ω is a bounded subset of $\mathbb{R}^n$ and m is n-dimensional Lebesgue measure, the notion of ubiquity was originally formulated by Dodson, Rynne & Vickers [18] to obtain lower bounds for the Hausdorff dimension of the sets $\Lambda(\psi)$ – see §5. Their notion of ubiquity is closely related to our notion of a 'local m-ubiquitous' system. In the case that the resonant sets are points the ubiquitous systems of Dodson, Rynne & Vickers coincide with the 'regular systems' of Baker & Schmidt [2]. Both these systems have proved very useful in obtaining lower bounds for the Hausdorff dimension of lim sup sets. However, both [2] and [18] fail to shed any light on the fundamental problems considered in this paper. For further details regarding regular systems and the original formulation of ubiquitous systems see [10].

Recently and independently, in [13] the notion of an optimal regular system introduced in [4] has been re-formulated to obtain divergent type Hausdorff measures results for subsets of $\mathbb{R}^n$. This re-formulated notion is essentially equivalent to our notion of local m-ubiquity in which m is n-dimensional Lebesgue measure, the resonant sets are points ($\gamma = 0$), the ubiquity function is comparable to $\rho : r \to r^{-1/n}$ and the sequences l and u are given by $l_{n+1} = u_n := 2^n$. These highly restrictive conditions, in particular the latter two which fix the function ρ and the sequences l and u, excludes many of the applications we have in mind even when Ω is a subset of $\mathbb{R}^n$. Furthermore, even with the restrictions our notion of local m-ubiquity is not equivalent to that of an optimal regular system since we make no assumption on the growth of $\#J_l^u(n)$.

3. The statements of the main theorems

First some useful notation. Let m be a measure satisfying condition (M1) with respect to the sequences l and u. Then $B_n(r)$ will denote a generic ball of radius r centred at a point of a resonant set R_α with α in $J_l^u(n)$. Given the conditions imposed on the measure m, we have that for any ball $B(c,r)$ with $c \in R_\alpha$ and $\alpha \in J_l^u(n)$

$$m(B(c,r)) \asymp m(B_n(r)) \,.$$

This comparability is obviously satisfied for any c in Ω if the measure satisfies (M2). With this in mind, we now state our main results. Recall, that an approximating function ψ is a real, positive decreasing function and that a ubiquity function ρ is a real, positive function such that $\rho(r) \to 0$ as $r \to \infty$.

THEOREM 1. *Let (Ω, d) be a compact metric space equipped with a measure m satisfying condition (M1) with respect to sequences l and u. Suppose that $(\mathcal{R}, \beta)$ is a global m-ubiquitous system relative to (ρ, l, u) and that ψ is an approximating function. Assume that*

$$(5) \qquad \limsup_{n \to \infty} \frac{\psi(u_n)}{\rho(u_n)} > 0$$

or assume that

$$(6) \qquad \sum_{n=1}^{\infty} \frac{m(B_n(\psi(u_n)))}{m(B_n(\rho(u_n)))} \left(\frac{\rho(u_n)}{\psi(u_n)} \right)^\gamma = \infty$$

and for Q sufficiently large

$$(7) \qquad \sum_{s=1}^{Q-1} \frac{\rho(u_s)^\gamma}{m(B_s(\rho(u_s)))} \sum_{\substack{s+1 \le t \le Q: \\ \psi(u_s) < \rho(u_t)}} \frac{m(B_t(\psi(u_t)))}{\psi(u_t)^\gamma} \ll \left(\sum_{n=1}^{Q} \frac{m(B_n(\psi(u_n)))}{m(B_n(\rho(u_n)))} \left(\frac{\rho(u_n)}{\psi(u_n)} \right)^\gamma \right)^2 .$$

Then, $m(\Lambda(\psi)) > 0$. In addition, if any open subset of Ω is m-measurable and $(\mathcal{R}, \beta)$ is locally m-ubiquitous relative to (ρ, l, u), then $m(\Lambda(\psi)) = 1$.

Before stating the Hausdorff measure analogue of the above theorems we introduce the following notion. Given a sequence u, a function h will be said to be **u-regular** if there exists a strictly positive constant $\lambda < 1$ such that for n sufficiently large

$$(8) \qquad h(u_{n+1}) \leq \lambda h(u_n) \ .$$

The constant λ is independent of n but may depend on u. Clearly, if h is u-regular then the function h is eventually, strictly decreasing along the sequence u. Thus, the regularity condition imposes the condition that u is eventually, strictly increasing. Also, note that if h is u-regular then it is s–regular for any subsequence s of u.

THEOREM 2. *Let (Ω, d) be a compact metric space equipped with a measure m satisfying condition* (M2). *Suppose that $(\mathcal{R}, \beta)$ is a locally m-ubiquitous system relative to (ρ, l, u) and that ψ is an approximation function. Let f be a dimension function such that $r^{-\delta} f(r) \to \infty$ as $r \to 0$ and $r^{-\delta} f(r)$ is decreasing. Furthermore, suppose that $r^{-\gamma} f(r)$ is increasing. Let g be the real, positive function given by*

$$(9) \qquad g(r) := f(\psi(r))\psi(r)^{-\gamma}\rho(r)^{\gamma-\delta} \quad \text{and let} \quad G := \limsup_{n \to \infty} g(u_n).$$

(i) *Suppose that $G = 0$ and that ρ is u-regular. Then,*

$$(10) \qquad \mathcal{H}^f(\Lambda(\psi)) = \infty \qquad \text{if} \qquad \sum_{n=1}^{\infty} g(u_n) = \infty \ .$$

(ii) *Suppose that $0 < G \leq \infty$. Then, $\mathcal{H}^f(\Lambda(\psi)) = \infty$.*

An important general observation: In statements such as Theorem 2 in which the measure is of type (M2), the lower sequence l is actually redundant from the hypothesis that $(\mathcal{R}, \beta)$ is a locally m-ubiquitous system relative to (ρ, l, u). The point is that the measure condition (M2) is independent of the sequences l and u. Hence, in view of the discussion in §2.4, given an upper sequence u for which

$$m\left(B \cap \bigcup_{\alpha \in J: \beta_\alpha \leq u_n} \Delta(R_\alpha, \rho(u_n))\right) \geq \kappa\, m(B) \ ,$$

a lower sequence l can always be constructed so that $(\mathcal{R}, \beta)$ is a locally m-ubiquitous system relative to (ρ, l, u).

In the statement of Theorem 1 the sequences l and u are determined by the measure condition (M1) as well as by (5) – (7) and it is important that we have ubiquity with respect to these sequences.

4. Remarks and corollaries to Theorem 1

Obviously the first conclusion of Theorem 1 is significantly weaker than the other — positive measure of $\Lambda(\psi)$ as opposed to full measure. However, in practice it is much easier to establish 'global ubiquity' than 'local ubiquity'. Moreover, for certain applications it is possible to establish the subsidiary result that $m(\Lambda(\psi))$ is either zero or one. For example, this is the case for the classical set of ψ-well approximable numbers – see Theorem 2.7 of [**22**] and indeed for the majority of applications considered in §12. Thus for such applications establishing $m(\Lambda(\psi)) > 0$ is enough to deduce the full measure result and in view of Theorem 1 'global ubiquity' is all that is required.

It will become evident during the course of establishing Theorem 1 that the intersection conditions associated with either form of ubiquity are only required for $\alpha \in J$ with $l_n < \beta_\alpha \le u_n$ rather than for $\alpha \in J$ with $\beta_\alpha \le u_n$. Also, for the positive measure statement of Theorem 1 the doubling property of the measure m is only required for balls centred at resonant sets rather than at arbitrary points of Ω.

It is easy to verify that the lim sup condition (5) implies the divergent sum condition (6). Thus, whenever (7) is satisfied the lim sup condition is redundant. At first glance, (7) may look like a horrendous condition. Nevertheless, we shall see that it is both natural and not particularly restrictive. For example, suppose throughout the following discussion that the measure m satisfies condition (M2). Then the divergent sum condition (6) becomes

$$(11) \qquad \sum_{n=1}^{\infty} \left(\frac{\psi(u_n)}{\rho(u_n)} \right)^{\delta-\gamma} = \infty \,,$$

and (7) simplifies to

$$(12) \qquad \sum_{s=1}^{Q-1} \rho(u_s)^{\gamma-\delta} \sum_{\substack{s+1 \le t \le Q: \\ \psi(u_s) < \rho(u_t)}} \psi(u_t)^{\delta-\gamma} \ll \left(\sum_{n=1}^{Q} \left(\frac{\psi(u_n)}{\rho(u_n)} \right)^{\delta-\gamma} \right)^2 .$$

Reiterating the earlier remark, trivially the lim sup condition (5) implies (11). Thus, whenever (12) is satisfied (5) is redundant since (11) and (12) together already imply the desired conclusions.

Consider for the moment the special case that $\gamma = \delta$. Trivially, both (11) and (12) are satsified. Thus, Theorem 1 reduces to:

COROLLARY 1. *Let (Ω, d) be a compact metric space equipped with a measure m satisfying condition (M2). Suppose that $(\mathcal{R}, \beta)$ is a global m-ubiquitous system relative to (ρ, l, u) and that ψ is an approximating function. If $\gamma = \delta$ then $m(\Lambda(\psi)) > 0$. In addition, if any open subset of Ω is m-measurable and $(\mathcal{R}, \beta)$ is locally m-ubiquitous relative to (ρ, l, u), then $m(\Lambda(\psi)) = 1$.*

Next, suppose that the approximating function ψ is u-regular. Then, for $t > s$ with s sufficiently large we have that

$$\psi(u_t) \le \lambda^{t-s}\, \psi(u_s)$$

for some $0 < \lambda < 1$. Without loss of generality, assume that $\delta - \gamma > 0$ – the case $\delta = \gamma$ is covered by the above corollary and since the measure m satisfies condition (M2) we have that $\gamma \le \dim \Omega = \delta$. Then for Q sufficiently large, the L.H.S. of (12) is

$$\ll \sum_{s=1}^{Q-1} \left(\frac{\psi(u_s)}{\rho(u_s)}\right)^{\delta-\gamma} \sum_{s < t \le Q} (\lambda^{\delta-\gamma})^{t-s} \ll \sum_{n=1}^{Q} \left(\frac{\psi(u_n)}{\rho(u_n)}\right)^{\delta-\gamma}.$$

This together with (11) implies that (12) is satisfied. We now consider the case that the ubiquity function ρ is u-regular. It is easily verified that for Q sufficiently large,

$$\text{L.H.S. of (12)} \quad \ll \quad \sum_{n=2}^{Q} \psi(u_n)^{\delta-\gamma} \sum_{m=1}^{n-1} \rho(u_m)^{\gamma-\delta}$$

$$\ll \sum_{n=2}^{Q} \left(\frac{\psi(u_n)}{\rho(u_n)}\right)^{\delta-\gamma} \sum_{m=1}^{n-1} (\lambda^{\delta-\gamma})^{n-m} \ll \sum_{n=1}^{Q} \left(\frac{\psi(u_n)}{\rho(u_n)}\right)^{\delta-\gamma}.$$

This together with (11) implies that (12) is satisfied. On gathering together these observations we have:

COROLLARY 2. *Let (Ω, d) be a compact metric space equipped with a measure m satisfying condition (M2). Suppose that $(\mathcal{R}, \beta)$ is a global m-ubiquitous system relative to (ρ, l, u) and that ψ is an approximating function. Furthermore, if $\delta > \gamma$ suppose that either ψ or ρ is u-regular and (11) is satisfied. Then $m(\Lambda(\psi)) > 0$. In addition, if any open subset of Ω is m-measurable and $(\mathcal{R}, \beta)$ is locally m-ubiquitous relative to (ρ, l, u), then $m(\Lambda(\psi)) = 1$.*

In the numerous applications considered in this paper, the various ubiquitous functions ρ will always satisfy (8) for the appropriate upper sequences u. Thus, if the measures are also of type (M2) then Theorem 1 simplifies to the above corollaries.

5. Remarks and corollaries to Theorem 2

The case when G is finite constitutes the main substance of the theorem. When $G = 0$, it would be desirable to remove the condition that the ubiquity function ρ is u-regular – see below; in particular Corollary 3. However, for the numerous applications considered in this paper the hypotheses that ρ is u-regular is always satisfied for the sequences u under consideration. Clearly, the assumption that the function $0 < G \le \infty$ in part (ii) implies the divergent sum condition in part (i).

The case when the dimension function f is δ–dimensional Hausdorff measure $\mathcal{H}^\delta$ is excluded from the statement of Theorem 2 by the condition that $r^{-\delta} f(r) \to \infty$ as $r \to 0$. This is natural since otherwise Theorem 1 implies that $m(\Lambda(\psi)) > 0$ which in turn implies that $\mathcal{H}^\delta(\Lambda(\psi))$ is positive and finite – see §7. In other words $\mathcal{H}^\delta(\Lambda(\psi))$ is never infinite. To see that Theorem 1 is applicable, note that with $f(r) = r^\delta$ the function g of Theorem 2 becomes $g(r) = (\psi(r)/\rho(r))^{\delta-\gamma}$. Thus, if the sum in Theorem 2 diverges then so does the sum in Theorem 1; i.e. (6) is satisfied. Now, if $G = 0$ then (7) is satisfied since we assume that ρ is u-regular in part (i) of Theorem 2. On the other hand, if $G > 0$ then (5) is satisfied. Thus, in either case Theorem 1 implies that $m(\Lambda(\psi)) > 0$.

Next notice that if $\gamma = 0$, then the hypothesis that $r^{-\gamma} f(r)$ is increasing is redundant since, by definition, any dimension function f is increasing (cf. §7). Also, notice that the conditions on $r^{-\gamma} f(r)$ and $r^{-\delta} f(r)$ exclude the possibility that $\gamma = \delta$. However, this is no great loss since if $\gamma = \delta$ then Corollary 1 implies that $m(\Lambda(\psi)) > 0$. Here we make use of the fact that 'local' implies global' ubiquity. Thus, $\mathcal{H}^\delta(\Lambda(\psi)) > 0$ and is in fact finite. Now, f is a dimension function such that $r^{-\delta} f(r) \to \infty$ as $r \to 0$. It is therefore a simple consequence of the elementary fact stated in §7 that $\mathcal{H}^f(\Lambda(\psi)) = \infty$. Thus without loss of generality we can assume that $\gamma < \delta$ in Theorem 2. Finally, note that in the case $\gamma > 0$, if both the functions ρ and g are decreasing then one always has that $r^{-\gamma} f(r)$ is increasing.

As mentioned above, when $G = 0$ it would be desirable to remove the condition that the ubiquity function ρ is u-regular in Theorem 2. At the expense of imposing growth conditions on the functions ψ and f, the following result achieves precisely this.

COROLLARY 3. *Let (Ω, d) be a compact metric space equipped with a measure m satisfying condition (M2). Suppose that $(\mathcal{R}, \beta)$ is a locally m-ubiquitous system relative to (ρ, l, u) and that ψ is an approximation function. Let f be a dimension function such that $r^{-\delta} f(r) \to \infty$ as $r \to 0$ and $r^{-\delta} f(r)$ is decreasing. Furthermore, suppose that $r^{-\gamma} f(r)$ is increasing. Let g be the positive function given by (9).*

(i) *Suppose that $G = 0$ and that ψ is u-regular. Furthermore, suppose there exist constants $r_0, \lambda_1, \lambda_2 \in (0,1)$ so that for any $r \in (0, r_0)$ one has $r^{\gamma} f(\lambda_1 r) \le \lambda_2 f(r)(\lambda_1 r)^{\gamma}$. Then, (10) is satisfied.*

(ii) *Suppose that $0 < G \le \infty$. Then, $\mathcal{H}^f(\Lambda(\psi)) = \infty$.*

Proof of Corollary 3. Recall that if the hypotheses of local ubiquity are satisfied for a particular upper sequence u then they are also satisfied for any subsequence s of u. The corollary follows from Theorem 2 by proving the existence of an appropriate subsequence s of u on which ρ is s-regular and $\sum g(s_n) = \infty$. To this end, since ψ is u-regular there exists a constant $\lambda \in (0,1)$ such that $\psi(u_{n+1}) \le \lambda \psi(u_n)$ for all n sufficiently large. Without loss of generality, we can assume that $\lambda \le \lambda_1$ – see §10.1.2. In view of the growth condition imposed on the dimension function f and the fact that $r^{-\gamma} f(r)$ is increasing, we have that for n sufficiently large

$$x_{n+1} = \frac{f(\psi(u_{n+1}))}{\psi(u_{n+1})^{\gamma}} \le \frac{f(\lambda \psi(u_n))}{\lambda \psi(u_n)^{\gamma}} \le \frac{f(\lambda_1 \psi(u_n))}{\lambda_1 \psi(u_n)^{\gamma}} \le \lambda_2 \frac{f(\psi(u_n))}{\psi(u_n)^{\gamma}}.$$

Hence, $x_{n+1} \le \lambda_2 x_n$. Next, fix some sufficiently large n_1 and for $k \ge 2$ let n_k be the least integer strictly greater than n_{k-1} such that $\rho(u_{n_k}) \le \frac{1}{2}\rho(u_{n_{k-1}})$. This is possible since $\rho(r) \to 0$ as $r \to \infty$. By construction, $\rho(u_m) \ge \frac{1}{2}\rho(u_{n_{k-1}})$ for any integer $m \in [n_{k-1}, n_k - 1]$. For shortness, let $\delta' := \delta - \gamma$. It follows that

$$
\begin{aligned}
\infty &= \sum_{n=n_1}^{\infty} g(u_n) = \sum_{n=n_1}^{\infty} x_n \, \rho(u_n)^{-\delta'} = \sum_{k=2}^{\infty} \sum_{n_{k-1} \le m < n_k} x_m \, \rho(u_m)^{-\delta'} \\
&\le \sum_{k=2}^{\infty} \sum_{n_{k-1} \le m < n_k} x_m \, \rho(u_{n_{k-1}})^{-\delta'} 2^{\delta'} = 2^{\delta'} \sum_{k=2}^{\infty} \rho(u_{n_{k-1}})^{-\delta'} \sum_{n_{k-1} \le m < n_k} x_m \\
&\ll \sum_{k=2}^{\infty} \rho(u_{n_{k-1}})^{-\delta'} x_{n_{k-1}} \sum_{i=0}^{\infty} \lambda_2^i \ll \sum_{k=1}^{\infty} \rho(u_{n_k})^{-\delta'} x_{n_k} := \sum_{k=1}^{\infty} g(u_{n_k}).
\end{aligned}
$$

Now set $s := \{u_{n_k}\}$. By construction, ρ is s-regular and $\sum g(s_n) = \infty$. #

Regarding Corollary 3, the growth condition imposed on the function f is not particularly restrictive. In particular, when $f : r \to r^s$ and $s > \gamma$ the growth condition is trivially satisfied. By restricting our attention to s-dimensional Hausdorff measure $\mathcal{H}^s$, Theorem 2 together with Corollary 3 yield the following statement.

COROLLARY 4. *Let (Ω, d) be a compact metric space equipped with a measure m satisfying condition* (M2). *Suppose that $(\mathcal{R}, \beta)$ is a locally m-ubiquitous system relative to (ρ, l, u) and that ψ is an approximation function. Let $s \geq 0$ such that $\gamma < s < \delta$, $g(r) := \psi(r)^{s-\gamma} \rho(r)^{\gamma - \delta}$ and let $G := \limsup_{n \to \infty} g(u_n)$.*

(i) *Suppose that $G = 0$ and that either ψ or ρ is u-regular. Then,*

$$\mathcal{H}^s(\Lambda(\psi)) \; = \; \infty \qquad \text{if} \qquad \sum_{n=1}^{\infty} g(u_n) \; = \; \infty \;\; .$$

(ii) *Suppose that $0 < G \leq \infty$. Then, $\mathcal{H}^s(\Lambda(\psi)) \; = \; \infty$.*

The following lower bound statement for the dimension of $\Lambda(\psi)$ is essentially a consequence of part (ii) of Corollary 4. The statement is free of any regularity condition.

COROLLARY 5. *Let (Ω, d) be a compact metric space with a measure m satisfying condition* (M2). *Suppose that $(\mathcal{R}, \beta)$ is a local m-ubiquitous system relative to (ρ, l, u) and let ψ be an approximating function.*

(i) *If $\delta \neq \gamma$ and $\lim_{n \to \infty} \psi(u_n)/\rho(u_n) = 0$ then*

$$\dim \Lambda(\psi) \geq d := \gamma + \sigma(\delta - \gamma), \quad \text{where } \sigma := \limsup_{n \to \infty} \frac{\log \rho(u_n)}{\log \psi(u_n)}.$$

Moreover, if $\liminf_{n \to \infty} \rho(u_n)/\psi(u_n)^\sigma < \infty$, then $\mathcal{H}^d(\Lambda(\psi)) = \infty$.

(ii) *If either $\delta = \gamma$ or $\limsup_{n \to \infty} \psi(u_n)/\rho(u_n) > 0$ then $0 < \mathcal{H}^\delta(\Lambda(\psi)) < \infty$ and so $\dim \Lambda(\psi) = \delta$.*

It will be evident from the proof below that part (ii) of Corollary 5 is a simple consequence of Theorem 1. Part (i) contains the main substance of Corollary 5. In order to establish part (i), all that is required of Corollary 4 is part (ii) and this explains why there is no regularity condition on either ψ or ρ in the statement of Corollary

5. Note that Corollary 4 implies both the dimension and measure statements of part (i) and moreover provides a more general criteria for when $\mathcal{H}^d(\Lambda(\psi)) = \infty$. To illustrate this, consider the classical set $W(\psi)$ of ψ–well approximable numbers with $\psi : r \to r^{-\tau}(\log r)^{-1}$ and $\tau > 2$ – see §1.1. The associated ubiquitous system is given by Lemma 1 in §2 – in particular $\rho(r) := \text{constant} \times r^{-2}$. Thus, $\sigma = 2/\tau$ and part (i) of Corollary 5 implies that $\dim W(\tau) \geq 2/\tau$ for $\tau > 2$. However, $\lim_{n \to \infty} \rho(2^n)/\psi(2^n)^\sigma = \infty$ and so we obtain no information concerning $\mathcal{H}^{2/\tau}(\Lambda(\psi))$. On the other hand, since the function ρ is u-regular Corollary 4 implies the above dimension statement and shows that $\mathcal{H}^{2/\tau}(\Lambda(\psi)) = \infty$. The dimension statement follows from the definition of Hausdorff dimension – see §7.

Proof Corollary 5. To start with consider the case $\delta = \gamma$, so $d := \gamma + \sigma(\delta - \gamma) = \delta$. Since 'local' implies 'global' ubiquity, Corollary 1 of Theorem 1 implies that $m(\Lambda(\psi)) > 0$. In turn, Lemma 5 of §7 implies that $0 < \mathcal{H}^\delta(\Lambda(\psi)) < \infty$ and $\dim \Lambda(\psi) = \delta$. This completes the proof in the case that $\delta = \gamma$. Next suppose, $\limsup_{n \to \infty} \psi(u_n)/\rho(u_n) > 0$. Then, Theorem 1 implies that $m(\Lambda(\psi)) > 0$ and so $0 < \mathcal{H}^\delta(\Lambda(\psi)) < \infty$ and $\dim \Lambda(\psi) = \delta$ for the same reasons as in the case $\delta = \gamma$. The completes the proof of part (ii) of Corollary 5. Thus, without loss of generality we can assume that $\lim_{n \to \infty} \psi(u_n)/\rho(u_n) = 0$ and so

$$0 \leq \limsup_{n \to \infty} \frac{\log \rho(u_n)}{\log \psi(u_n)} := \sigma \leq 1 .$$

Regarding part (i), we first suppose that there exists a strictly increasing sequence $\{n_i\}_{i \in \mathbb{N}}$ such that

$$(13) \qquad \lim_{i \to \infty} \frac{\rho(u_{n_i})}{\psi(u_{n_i})^\sigma} = L < \infty .$$

Since $\lim_{n \to \infty} \psi(u_n)/\rho(u_n) = 0$ we have that $\sigma < 1$. This together with the fact that $\delta > \gamma$ implies that $\gamma \leq d < \delta$. Now notice that (13) implies that

$$\lim_{i \to \infty} \frac{\psi(u_{n_i})^{d - \gamma}}{\rho(u_{n_i})^{\delta - \gamma}} := \lim_{i \to \infty} g(u_{n_i}) = L^{\gamma - \delta} > 0 ,$$

and g is precisely the function in Corollary 4 with $s = d$. Hence $G > 0$ and part (ii) of Corollary 4 implies that $\mathcal{H}^d(\Lambda(\psi)) = \infty$, as required. Now suppose there is no sequence $\{n_i\}$ such that (13) is satisfied. Then $\sigma > 0$ since $\rho(r) \to 0$ as $r \to \infty$. It follows from the definition of σ that for any $0 < \epsilon < \sigma$, there exists a sequence $\{n_i\}_{i \in \mathbb{N}}$ such that (13) is satisfied with σ replaced by $\sigma_\epsilon := \sigma - \epsilon$. In fact, $L = 0$ in (13). Thus, on repeating the above argument with σ replaced by σ_ϵ

and d replaced by $d_\epsilon := \gamma + \sigma_\epsilon(\delta - \gamma)$ we conclude that $\mathcal{H}^{d_\epsilon}(\Lambda(\psi)) = \infty$. Hence $\dim \Lambda(\psi) \geq d_\epsilon = d - \epsilon(\delta - \gamma)$. On letting $\epsilon \to 0$, we obtain the desired dimension result. This completes the proof of Corollary 5.   #

The following corollary is a simple consequence of Corollary 5. In some sense it is no more than a slightly weaker, alternative statement and is more in line with the original 'ubiquity result' of Dodson, Rynne & Vickers [18].

COROLLARY 6. *Let (Ω, d) be a compact metric space equipped with a measure m satisfying condition* (M2). *Suppose $(\mathcal{R}, \beta)$ is a local m-ubiquitous system relative to (ρ, l, u) and let ψ be an approximating function. Then $\dim \Lambda(\psi) \geq d := \gamma + \sigma(\delta - \gamma)$, where*

$$\sigma := \min\left\{1, \left|\limsup_{n \to \infty} \frac{\log \rho(u_n)}{\log \psi(u_n)}\right|\right\}.$$

Furthermore, if $d < \delta$ and $\liminf_{n \to \infty} \rho(u_n)/\psi(u_n)^\sigma < \infty$, then $\mathcal{H}^d(\Lambda(\psi)) = \infty$.

In the case that (Ω, d) is a bounded subset of $\mathbb{R}^n$, m is the Lebesgue measure in $\mathbb{R}^n$ and $\liminf_{n \to \infty} \rho(u_n)/\psi(u_n)^\sigma = 0$, the above corollary is essentially the 'ubiquity result' of Dodson, Rynne & Vickers [18][3]. Also, in the case that (Ω, d) is a bounded subset of $\mathbb{R}^n$, the resonant sets are points ($\gamma = 0$) and $\liminf_{n \to \infty} \rho(u_n)/\psi(u_n)^\sigma = 0$, the above corollary is essentially equivalent to Theorem 1 of [17].

We end this section with a comment regarding Theorem 2 and the measure condition (M2). Given the central conditions on m (namely that the m–measure of a ball $B(x, r)$ with $r > 0$ and $x \in \Omega$ is strictly positive and that m is doubling) the main property of (M2) that is utilized during the proof of Theorem 2 is that $m(B(x, r))$ is comparable to a function of r alone and is independent of x. In view of this, consider the following measure condition.

(M2′)   There exists a positive constant r_o such that for any $x \in \Omega$ and $r \leq r_o$,

$$a\,\mathbf{m}(r) \ \leq \ m(B(x, r)) \ \leq \ b\,\mathbf{m}(r)$$

where $\mathbf{m} : \mathbb{R}^+ \to \mathbb{R}^+$ is an increasing, continuous function with $\mathbf{m}(r) \to 0$ as $r \to 0$.

---

[3]The statement of the Theorem 1 in [18] is not correct when $\delta = \gamma$ and the proof assumes that $\lim_{r \to \infty} \psi(r) = 0$. Also, a weaker statement of Corollary 6 appears as Theorem 5.6 in [10]. However, the proof contains a flaw. The claim that a certain set T_∞ is a subset of $\Lambda(\psi)$ is not necessarily true.

Clearly, with $\mathbf{m}(r) := r^\delta$ we obtain the measure condition (M2). By adapting the proof of Theorem 2 in the obvious manner we obtain the more general result:

THEOREM 2′ *Let (Ω, d) be a compact metric space equipped with a measure m satisfying condition (M2′). Suppose that $(\mathcal{R}, \beta)$ is a locally m-ubiquitous system relative to (ρ, l, u) and that ψ is an approximation function. Let f be a dimension function such that $f(r)/\mathbf{m}(r) \to \infty$ as $r \to 0$ and $f(r)/\mathbf{m}(r)$ is decreasing. Furthermore, suppose that $r^{-\gamma}\mathbf{m}(r)$ is increasing and that $r^{-\gamma} f(r)$ is increasing. Let g be the function given by*

$$g(r) := f(\psi(r)) \, \psi(r)^{-\gamma} \rho(r)^\gamma \, \mathbf{m}(\rho(r))^{-1} \quad \text{and let} \quad G := \limsup_{r \to \infty} g(u_n).$$

(i) Suppose that $G = 0$ and that ρ is u-regular. Then,

$$\mathcal{H}^f(\Lambda(\psi)) \;=\; \infty \qquad \text{if} \qquad \sum_{n=1}^{\infty} g(u_n) \;=\; \infty \;\; .$$

(ii) Suppose that $0 < G \leq \infty$. Then, $\mathcal{H}^f(\Lambda(\psi)) = \infty$.

We have opted to work with condition (M2) rather than (M2′) simply for the sake of clarity and the ease of discussion. Furthermore, for the various applications considered in §12 if m satisfies (M2′) then it always satisfies (M2). Note that the extra condition that $r^{-\gamma}\mathbf{m}(r)$ is increasing in Theorem 2′ plays the role of the fact that $\gamma \leq \delta$ in the case that $\mathbf{m}(r) := r^\delta$.

6. The classical results

For the classical set $W(\psi)$ of ψ–well approximable numbers, Lemma 1 in §2 establishes local m-ubiquity. Clearly, the ubiquity function ρ satisfies (8) (i.e. ρ is u-regular) and so Corollary 2 establishes the divergent part of Khintchine's Theorem. In fact global m-ubiquity would suffice since to go from positive measure to full measure simply involves making use of the subsidiary result that $m(\Lambda(\psi))$ is either zero or one - see Theorem 2.7 of [22]. On the other hand, Theorem 2 establishes the divergent part of Jarník's Theorem. By making use of the 'natural cover' of $W(\psi)$, the convergent parts of these classical results are easily established.

7. Hausdorff measures and dimension

A *dimension function* $f : \mathbb{R}^+ \to \mathbb{R}^+$ is an increasing, continuous function such that $f(r) \to 0$ as $r \to 0$. The Hausdorff f–measure with respect to the dimension function f will be denoted throughout by $\mathcal{H}^f$ and is defined as follows. Suppose F is a non–empty subset of (Ω, d). For $\rho > 0$, a countable collection $\{B_i\}$ of balls in Ω with radii $r_i \leq \rho$ for each i such that $F \subset \bigcup_i B_i$ is called a ρ-*cover* for F. Clearly such a cover always exists for totally bounded metric spaces. For a dimension function f define $\mathcal{H}_\rho^f(F) = \inf \left\{ \sum_i f(r_i) : \{B_i\} \text{ is a } \rho\text{–cover of } F \right\}$, where the infimum is over all ρ-covers. The *Hausdorff f–measure* $\mathcal{H}^f(F)$ of F with respect to the dimension function f is defined by

$$\mathcal{H}^f(F) := \lim_{\rho \to 0} \mathcal{H}_\rho^f(F) \ = \ \sup_{\rho > 0} \mathcal{H}_\rho^f(F) \ .$$

A simple consequence of the definition of $\mathcal{H}^f$ is the following useful

LEMMA 2. *If f and g are two dimension functions such that the ratio $f(r)/g(r) \to 0$ as $r \to 0$, then $\mathcal{H}^f(F) = 0$ whenever $\mathcal{H}^g(F) < \infty$.*

In the case that $f(r) = r^s$ $(s \geq 0)$, the measure $\mathcal{H}^f$ is the usual s–*dimensional Hausdorff measure* $\mathcal{H}^s$ and the Hausdorff dimension $\dim F$ of a set F is defined by

$$\dim F := \inf \{s : \mathcal{H}^s(F) = 0\} = \sup \{s : \mathcal{H}^s(F) = \infty\} \ .$$

In particular when s is an integer $\mathcal{H}^s$ is comparable to s–dimensional Lebesgue measure. For further details see [**16, 20**]. A general and classical method for obtaining a lower bound for the Hausdorff f-measure of an arbitrary set F is the following mass distribution principle.

LEMMA 3 (Mass Distribution Principle). *Let μ be a probability measure supported on a subset F of (Ω, d). Suppose there are positive constants c and r_o such that for any ball B with radius $r \leq r_o$*

$$\mu(B) \leq c \, f(r) \ .$$

If X is a subset of F with $\mu(X) = \lambda > 0$ then $\mathcal{H}^f(X) \geq \lambda/c$.

Proof. If $\{B_i\}$ is a ρ–cover of X with $\rho \leq r_o$ then

$$\lambda = \mu(X) = \mu\left(\bigcup_i B_i\right) \leq \sum_i \mu(B_i) \leq c\sum_i f(r_i) .$$

It follows that $\mathcal{H}^f_\rho(X) \geq \lambda/c$ for any $\rho \leq r_o$. On letting $\rho \to 0$, the quantity $\mathcal{H}^f_\rho(X)$ increases and so we obtain the required result.

$\#$

The following rather simple covering result will be used at various stages during the proof of our theorems.

LEMMA 4 (Covering lemma). *Let (Ω, d) be a metric space and $\mathcal{B}$ be a finite collection of balls with common radius $r > 0$. Then there exists a disjoint sub-collection $\{B_i\}$ such that*

$$\bigcup_{B \in \mathcal{B}} B \subset \bigcup_i 3B_i .$$

Proof. Let S denote the set of centres of the balls in $\mathcal{B}$. Choose $c_1 \in S$ and for $k \geq 1$,

$$c_{k+1} \in S \setminus \bigcup_{i=1}^{k} B(c_i, 2r)$$

as long as $S \setminus \bigcup_{i=1}^{k} B(c_i, 2r) \neq \emptyset$. Since $\#S$ is finite, there exists $k_1 \leq \#S$ such that $S \subset \bigcup_{i=1}^{k_1} B(c_i, 2r)$. By construction, any ball $B(c, r)$ in the original collection $\mathcal{B}$ is contained in some ball $B(c_i, 3r)$ and since $d(c_i, c_j) > 2r$ the chosen balls $B(c_i, r)$ are clearly disjoint.

$\#$

We end this section by making use of the mass distribution principle and the covering lemma to establish the following claim mentioned in §2.

LEMMA 5. *Let (Ω, d) be a totally bounded metric space equipped with a probability measure m satisfying condition (M2). Then for any $X \subseteq \Omega$ with $m(X) > 0$*

$$0 < \mathcal{H}^\delta(X) < \infty \qquad \text{and} \qquad \dim X = \delta .$$

Proof. Given the measure statement, the dimension statement follows directly from the definition of Hausdorff dimension. The fact that $\mathcal{H}^\delta(X)$ is strictly positive is a simple consequence of the mass distribution principle with $f(r) = r^\delta$. Thus the lemma follows on showing that $\mathcal{H}^\delta(\Omega)$ is finite since $\mathcal{H}^\delta(X) \leq \mathcal{H}^\delta(\Omega)$. Since the metric space is totally bounded, for any $\rho > 0$ there exists a finite collection $\mathcal{B}$ of balls $B(\rho)$ with centres in Ω and common radius ρ such that

$$\Omega \subset \bigcup_{B(\rho) \in \mathcal{B}} B(\rho) \, .$$

In other words, $\mathcal{B}$ is a ρ–cover of Ω. By the covering lemma, there exists a sub-collection $\{B_i(\rho)\}$ such that

$$\overset{\circ}{\bigcup_i} B_i(\rho) \subset \bigcup_{B(\rho) \in \mathcal{B}} B(\rho) \subset \bigcup_i B_i(3\rho) \, ,$$

where the left hand union is disjoint. Thus the collection $\{B_i(3\rho)\}$ is a 3ρ-cover of Ω. Hence,

$$\mathcal{H}^\delta_{3\rho}(\Omega) \leq \sum_i (3\rho)^\delta \ll \sum_i m(B_i(\rho)) = m\left(\overset{\circ}{\bigcup_i} B_i(\rho)\right) \leq m(\Omega) = 1.$$

On letting $\rho \to 0$, we obtain that $\mathcal{H}^\delta(\Omega) \ll 1$ as required.

$\#$

8. Positive and full m–measure sets

The aim of this section is to determine conditions under which a subset E of a metric measure space (Ω, d, m) has full m-measure. We also state two important lemmas which enable us to conclude that $m(E)$ is strictly positive in the case that E is a lim sup set. It is worth mentioning that the lemmas are in fact a generalization of results known for Lebesgue measure since the 1920s – see, for instance [31].

LEMMA 6. *Let (Ω, d) be a metric space and let m be a finite measure on Ω such that any open set is m-measurable. Let E be a Borel subset of Ω and $f : \mathbb{R}^+ \to \mathbb{R}^+$ be increasing with $f(x) \to 0$ as $x \to 0$. Assume that*

$$m(E \cap U) \geq f(m(U)) \quad \text{for any open set } U \subset \Omega.$$

Then E has full measure in Ω; i.e. $m(\Omega \setminus E) = 0$.

Proof. Assume that $m(\Omega \setminus E) > 0$. Thus there exists $\varepsilon > 0$ such that $f(m(\Omega \setminus E)) > \varepsilon$. Notice that $m(E) \leq m(\Omega) < \infty$. Then E contains a closed set C for which $m(E \setminus C) < \varepsilon$ (see Theorem 2.2.2. [21]). Since C is closed, the set $U = \Omega \setminus C$ is open and thus

$m(U \cap E) \geq f(m(U))$. As $C \subset E$, we have $\Omega \setminus E \subset U$. Then $m(\Omega \setminus E) \leq m(U)$ and by the monotonicity of f,

$$m(U \cap E) \geq f(m(U)) \geq f(m(\Omega \setminus E)) > \varepsilon.$$

However, $E \setminus C = U \cap E$. So the previous set of inequalities contradicts the fact that $m(U \cap E) = m(E \setminus C) < \varepsilon$.

$\#$

LEMMA 7. *Let (Ω, d) be a metric space and let m be a finite measure on Ω such that any open set is m-measurable. Let E be a Borel subset of Ω. Assume that there are constants $r_0, a > 0$ such that for any ball B with centre in Ω of radius $r(B) < r_0$*

$$m(E \cap B) \geq a\, m(B)\ .$$

Furthermore, assume that there is a constant $b > 0$ such that for every open set U there is a finite or countable disjoint collection G of balls contained in U with centres in Ω and radii less than r_0 such that

(14)
$$b\, m(U) \leq \sum_{B \in G} m(B).$$

Then E has full measure in Ω; i.e. $m(\Omega \setminus E) = 0$.

Proof. Take any open set U. Then there is a disjoint collection G of balls in U with centres in Ω and radii less than r_0 satisfying (14). It follows that $m(U \cap E) \geq m\left(\bigcup_{B \in G} B \cap E\right) = \sum_{B \in G} m(B \cap E) \geq \sum_{B \in G} a\, m(B) \geq a\, b\, m(U)$. Applying Lemma 6 with $f(x) = ab\, x$ completes the proof.

$\#$

In short, condition (14) implies that any open set U can be substantially packed with sufficiently small disjoint balls centred in Ω.

At this point we introduce various notions which can all be found in §2.8 of [**21**]. Let F be a family of closed subsets of Ω. We say that F covers $A \subset \Omega$ *finely* if for any $a \in A$, $\varepsilon > 0$ there is a set S in F such that $a \in S$ with S contained in the open ball $B(a, \varepsilon)$. Next, F is said to be *m-adequate* for A if for each open subset V of Ω there is a countable disjoint subfamily G of F with

$$\bigcup_{S \in G} S \subset V \text{ and } m\left((V \cap A) \setminus \bigcup_{S \in G} S\right) = 0.$$

Finally, we associate with each subset S of F its (δ, τ)-*enlargement* defined by

$$(15) \qquad\qquad \hat{S} = \bigcup_{T \in F,\ T \cap S \neq \emptyset,\ \delta(T) \leq \tau\, \delta(S)} T\,.$$

Here $\tau \in \mathbb{R}^+$ and δ is a non-negative bounded function on F. The following lemma brings together the above notions and appears as Theorem 2.8.7 in [**21**].

LEMMA ON ADEQUATE FAMILIES OF SETS *Assume that m is a finite measure on Ω such that every open set in Ω is m-measurable. If F covers $A \subset \Omega$ finely, δ is a non-negative bounded function on F, $1 < \tau, \lambda < \infty$ and*

$$(16) \qquad\qquad m(\hat{S}) \leq \lambda\, m(S)$$

whenever $S \in F$ and $\hat{S}$ is the (δ, τ)-enlargement of S, then F is m-adequate for A.

We now consider the case that F is the family of closed balls B with centres in Ω. With reference to Lemma 7, we assume that the radius $r(B)$ of any ball in F is less than r_o. It follows that F covers finely every subset of Ω and in particular any open subset U. Next, consider the function $\delta : B \to \delta(B)$ where $\delta(B)$ is the diameter of B in F. Then $S = B \in F$ implies $\hat{S} \subset (1 + 2\tau)\, B$. Thus, if the measure m satisfies the *diametric regularity condition*:

$$(17) \qquad\qquad m((1 + 2\tau)\, B) \ \leq \ \lambda\, m(B)\,,$$

then (16) is satisfied. Clearly, if m is doubling then the diametric regularity condition (17) is trivially satisfied. Recall, that m is said to be doubling if there exists a constant $C \geq 1$ such that $\forall\, x \in \Omega \quad m(B(x, 2r)) \leq C\, m(B(x, r))$. Note that the doubling condition is independent of whether the ball $B(x, r)$ is open or closed. It is easy to see that if m is doubling then the m measure of a closed ball $B(x, r)$ with x in Ω is comparable to that of the open ball $B^\circ(x, r)$:

$$\tfrac{1}{C}\, m(B) \leq m(\tfrac{1}{2} B) \leq m(B^\circ) \leq m(B)\,.$$

The upshot of this is that if F and δ are as above and m is doubling, then F is m-adequate for any open subset U of Ω. By definition, there is a finite or countable disjoint family $G \subset F$ of closed balls with radii less than r_o such that

$$\bigcup_{B \in G} B \subset U \quad \text{and} \quad m\Big(U \setminus \bigcup_{B \in G} B\Big) = 0.$$

Hence

$$m(U) = \sum_{B \in G} m(B).$$

Thus the family G of balls satisfies (14) and we obtain the following useful modification of Lemma 7.

PROPOSITION 1. *Let (Ω, d) be a metric space and let m be a finite, doubling measure on Ω such that any open set is measurable. Let E be a Borel subset of Ω. Assume that there are constants $r_0, c > 0$ such that for any ball B of radius $r(B) < r_0$ and centre in Ω we have that*

$$m(E \cap B) \geq c\, m(B) .$$

Then E has full measure in Ω, i.e. $m(\Omega \setminus E) = 0$.

Proposition 1 will be used in the proof of Theorem 1 to show the full measure result. The following two propositions on the m–measure of $\limsup$ sets will be required to establish that $m(\Lambda(\psi)) > 0$.

PROPOSITION 2. *Let (Ω, A, m) be a probability space and $E_n \in A$ be a sequence of sets such that $\sum_{n=1}^{\infty} m(E_n) = \infty$. Then*

$$m(\limsup_{n \to \infty} E_n) \;\geq\; \limsup_{Q \to \infty} \frac{\left(\sum_{s=1}^{Q} m(E_s) \right)^2}{\sum_{s,t=1}^{Q} m(E_s \cap E_t)} \;.$$

This result is a generalization of the divergent part of the standard Borel–Cantelli lemma. For the proof see either [**22**, Lemma 2.3] or [**39**, Lemma 5].

PROPOSITION 3. *Let (Ω, A, m) be a probability space, $F \in A$ and $E_n \in A$ a sequence of sets. Suppose there exists a constant $c > 0$ such that $\limsup_{n \to \infty} m(F \cap E_n) \geq c\, m(F)$. Then*

$$m(F \cap \limsup_{n \to \infty} E_n) \;\geq\; c^2\, m(F) \;.$$

Proof. Without loss of generality assume that $m(F) > 0$. For any $0 < \varepsilon < c$, there is a subsequence E_{n_i} with n_i strictly increasing such that $m(F \cap E_{n_i}) \geq (c - \varepsilon)\, m(F)$.

Clearly

$$\Big(\sum_{i=1}^{N} m(F \cap E_{n_i}) \Big)^2 \geq \Big(\sum_{i=1}^{N} (c - \varepsilon)\, m(F) \Big)^2 = (c - \varepsilon)^2 N^2\, m(F)^2$$

and

$$\sum_{n,m=1}^{N} m(F \cap E_n \cap E_m) \leq \sum_{m,n=1}^{N} m(F) = m(F)\, N^2.$$

Also notice that $\sum_{i=1}^{\infty} m(F \cap E_{n_i}) \geq m(F) \sum_{i=1}^{\infty}(c - \varepsilon) = \infty$. Thus on applying Proposition 2 and observing that $F \cap \limsup_{n\to\infty} E_n \supseteq F \cap \limsup_{i\to\infty} E_{n_i}$ we have that

$$m\Big(F \cap \limsup_{n\to\infty} E_n \Big) \geq \limsup_{N\to\infty} \frac{(c - \varepsilon)^2 N^2 m(F)^2}{m(F) N^2} = (c - \varepsilon)^2\, m(F) \ .$$

As $\varepsilon > 0$ is arbitrary, this completes the proof of the proposition.   #

9. Proof of Theorem 1

Let B be an arbitrary ball centred at a point in Ω. The aim is to show that

(18) $$m(\Lambda(\psi) \cap B) \ \geq \ m(B)/C \ ,$$

where $C > 0$ is a constant independent of B.

Under the global ubiquity hypothesis, Theorem 1 follows on establishing (18) with $B := \Omega$ – the space Ω can be regarded as a ball since it is compact. In the case of local ubiquity, (18) will be established for balls B with sufficiently small radii so that the conditions of local ubiquity and Proposition 1 are fulfilled. Then (18) together with Proposition 1 clearly implies Theorem 1 for local ubiquity. Since in the 'local' case we appeal to Proposition 1, the extra hypothesis that any open subset of Ω is m-measurable is necessary.

In view of the above discussion, let $B(x, r)$ be a ball for which (4) is satisfied. In order to establish (18), we begin by constructing a 'good' subset $A(\psi, B)$ of $\Lambda(\psi) \cap B$. Essentially, each thickening $\Delta(R_\alpha, \psi(\beta_\alpha))$ of a resonant set R_α will be replaced by carefully chosen collections of balls contained in the set $\Delta(R_\alpha, \psi(\beta_\alpha))$ with centers on R_α. In the case that the resonant sets are points ($\gamma = 0$), so that the thickenings themselves are already balls, the argument is much simplified but still crucial.

9.1. The subset $A(\psi, B)$ **of** $\Lambda(\psi) \cap B$. Since (Ω, d) is totally bounded, we can cover Ω by a finite collection of balls $\tilde{B}$ with common radius $\rho(u_n)$. Suppose $\tilde{B} \cap \Delta(R_\alpha, \rho(u_n)) \neq \emptyset$. Then there exists a point $c \in R_\alpha$ such that $\tilde{B} \subset B(c, 3\rho(u_n))$. Thus, for each $\alpha \in J_l^u(n)$, there is a finite cover of $\Delta(R_\alpha, \rho(u_n))$ by balls $B(c, 3\rho(u_n))$ centred at points c on R_α. This statement is of course obvious in the case when the resonant sets are points. Denote by $G_\Omega(n, \alpha)$ the collection of centers $c \in R_\alpha$ of the balls $B(c, 3\rho(u_n))$ and by $G_\Omega^*(n)$ the set of all such centers as α runs through $J_l^u(n)$; that is $G_\Omega^*(n) := \{c \in G_\Omega(n, \alpha) : \alpha \in J_l^u(n)\}$. In the case that c lies on more than one R_α simply choose one of them. Clearly, the collection of balls $\mathcal{B}_\Omega^*(n) := \{B(c, 3\rho(u_n)) : c \in G_\Omega^*(n)\}$ is a cover for $\Delta_l^u(\rho, n) := \bigcup_{\alpha \in J_l^u(n)} \Delta(R_\alpha, \rho(u_n))$. In view of the covering lemma, there exists a disjoint sub-collection $\tilde{\mathcal{B}}_\Omega(n)$ of $\mathcal{B}_\Omega^*(n)$ with centers $c \in \tilde{G}_\Omega(n)$ such that

$$(19) \qquad \overset{\circ}{\bigcup_{c \in \tilde{G}_\Omega(n)}} B(c, \rho(u_n)) \subset \Delta_l^u(\rho, n) \subset \bigcup_{c \in \tilde{G}_\Omega(n)} B(c, 9\rho(u_n)) \ .$$

The left hand side follows from the fact that the balls $B(c, 3\rho(u_n))$ with $c \in \tilde{G}_\Omega(n)$ are disjoint and that $B(c, \rho(u_n)) \subseteq \Delta(R_\alpha, \rho(u_n))$ for any point c on R_α.

Choose n sufficiently large so that $36\rho(u_n) < r$ (by definition, $\rho(u_n) \to 0$ as $n \to \infty$) and let

$$G_B(n) := \left\{ c \in \tilde{G}_\Omega(n) : c \in \tfrac{1}{2}B \right\} \ .$$

Now by definition and (19),

$$\overset{\circ}{\bigcup_{c \in G_B(n)}} B(c, \rho(u_n)) \subset \Delta_l^u(\rho, n) \cap B$$

and

$$\bigcup_{c \in G_B(n)} B(c, 9\rho(u_n))) \supset \Delta_l^u(\rho, n) \cap \tfrac{1}{4}B \ .$$

We now estimate the cardinality of $G_B(n)$. By (4) and the fact that the measure m is of type (M1), for n sufficiently large

$$\#G_B(n) \, m(B_n(\rho(u_n))) \gg m\left(\bigcup_{c \in G_B(n)} B(c, 9\rho(u_n))\right) \geq$$
$$\geq m\left(\Delta_l^u(\rho, n) \cap \tfrac{1}{4}B\right) \geq \kappa \, m(\tfrac{1}{4}B) \gg \kappa \, m(B) \ .$$

where $B_n(r)$ is a generic ball centred at a point $c \in G_\Omega(n)$ of radius r. On the other hand

$$m(B) \;\geq\; m\left(\bigcup_{c \in G_B(n)}^{\circ} B(c, \rho(u_n)) \right) \;\gg\; \#G_B(n)\, m(B_n(\rho(u_n)))\;,$$

where the implied constant is dependent only on the constant a of (2). The upshot of this is that

$$(20) \qquad\qquad\qquad \#G_B(n) \;\asymp\; \frac{m(B)}{m(B_n(\rho(u_n)))} \;\;.$$

In the case $B = \Omega$, (20) is satisfied with $m(B)$ replaced by $m(\Omega) := 1$.

We are now already in the position to prove the theorem under the $\limsup$ hypothesis (5). Suppose for some sufficiently large $n \in \mathbb{N}$ we have that $\psi(u_n) \geq k\,\rho(u_n)$, where $k > 0$ is a constant. If $k \geq 1$, (4) implies that

$$m(\Delta_l^u(\psi, n) \cap B) \;\geq\; m(\Delta_l^u(\rho, n) \cap B) \;\geq\; \kappa\, m(B)\;.$$

On the other hand if $\rho(u_n) > \psi(u_n) > k\,\rho(u_n)$, then (20) together with the fact that m is doubling and that $k < 1$ implies that

$$\begin{aligned}
m(\Delta_l^u(\psi, n) \cap B) \;&\geq\; m\Big(\bigcup_{c \in G_B(n)}^{\circ} B(c, \psi(u_n)) \Big) \\
&\gg\; \#G_B(n)\, m\left(B_n(\psi(u_n)) \right) \;\gg\; m(B)\;.
\end{aligned}$$

Thus, if $\psi(u_n) \geq k\,\rho(u_n)$ for infinitely many $n \in \mathbb{N}$, Proposition 3 with $F = \Omega$ implies (18) and thereby completes the proof of Theorem 1. It remains to establish the theorem under the hypotheses (6) and (7). Moreover, given any constant $k > 0$ we can assume without loss of generality that for n sufficiently large

$$(21) \qquad\qquad\qquad\qquad \rho(u_n) > k\,\psi(u_n)\;.$$

By definition, for each $c \in G_B(n)$ there exists an $\alpha \in J_l^u(n)$ such that $c \in R_\alpha \cap \frac{1}{2}B$ or simply that $c \in R_\alpha$ when $B = \Omega$. Assume for the moment that $\gamma > 0$. Cover the set

$$B(c, \tfrac{1}{2}\rho(u_n)) \cap \Delta(R_\alpha, \psi(u_n)) \;\subset\; B$$

by balls $\tilde{B}$ of common radius $\psi(u_n)$. Suppose $\tilde{B} \cap B(c, \tfrac{1}{2}\rho(u_n)) \cap \Delta(R_\alpha, \psi(u_n)) \neq \emptyset$. Then there exists some $c' \in R_\alpha$ such that $\tilde{B} \subset B(c', 3\psi(u_n))$. Let $\mathcal{B}_B^*(n, c)$ denote the collection of balls $B(c', 3\psi(u_n))$ arising in this way. Clearly this collection of balls centred at points on R_α is a cover for the set $B(c, \tfrac{1}{2}\rho(u_n)) \cap \Delta(R_\alpha, \psi(u_n))$. Note that since we can assume that $\rho(u_n) > 24\,\psi(u_n)$ for n large enough, the collection $\mathcal{B}_{\mathrm{B}}^*(n,c)$ of balls is contained in $B(c, \frac{3}{4}\rho(u_n))$.

By the covering lemma, there is a disjoint sub-collection $\mathcal{B}_{\mathrm{B}}(n,c)$ of $\mathcal{B}_{\mathrm{B}}^*(n,c)$ with centers $c' \in G_{\mathrm{B}}(n,c)$ such that for n sufficiently large

$$(22) \qquad B(c, \tfrac{1}{2}\rho(u_n)) \cap \Delta(R_\alpha, \psi(u_n)) \subset \bigcup_{c' \in G_{\mathrm{B}}(n,c)} B(c', 9\psi(u_n)),$$

and

$$(23) \qquad \overset{\circ}{\bigcup_{c' \in G_{\mathrm{B}}(n,c)}} B(c', \psi(u_n)) \subset B(c, \tfrac{3}{4}\rho(u_n)) \cap \Delta(R_\alpha, \psi(u_n)) \ .$$

Now, (22) together with intersection condition (i), implies that

$$\#G_{\mathrm{B}}(n,c)\, m\left(B_n(\psi(u_n))\right) \ \gg \ m\left(\bigcup_{c' \in G_{\mathrm{B}}(n,c)} B(c', 9\psi(u_n))\right)$$

$$\geq \ m\left(B(c, \tfrac{1}{2}\rho(u_n)) \cap \Delta(R_\alpha, \psi(u_n))\right)$$

$$\gg \ m\left(B_n(\psi(u_n))\right) \times \left(\frac{\rho(u_n)}{\psi(u_n)}\right)^\gamma .$$

Similarly, (23) together with intersection condition (ii) implies that

$$\#G_{\mathrm{B}}(n,c)\, m(B_n(\psi(u_n))) \ \asymp \ m\left(\overset{\circ}{\bigcup_{c' \in G_{\mathrm{B}}(n,c)}} B(c', \psi(u_n))\right)$$

$$\leq \ m\left(B(c, \tfrac{3}{4}\rho(u_n)) \cap \Delta(R_\alpha, \psi(u_n))\right)$$

$$\ll \ m(B_n(\psi(u_n))) \times \left(\frac{\rho(u_n)}{\psi(u_n)}\right)^\gamma .$$

Hence,

$$(24) \qquad\qquad \#G_{\mathrm{B}}(n,c) \ \asymp \ \left(\frac{\rho(u_n)}{\psi(u_n)}\right)^\gamma \ .$$

In the case $\gamma = 0$, we define $G_{\mathrm{B}}(n,c) := \{c\}$ and so $\#G_{\mathrm{B}}(n,c) = 1$. Thus (24) is satisfied even when $\gamma = 0$. Now let

$$A_n(\psi, B) := \bigcup_{c \in G_{\mathrm{B}}(n)} \bigcup_{c' \in G_{\mathrm{B}}(n,c)} B(c', \psi(u_n)) \ .$$

It is easily verified that the balls in the above definition of $A_n(\psi, B)$ are disjoint. Indeed, for any $c \in G_B(n)$ the balls $B(c', \psi(u_n))$ with $c' \in G_B(n, c)$ are disjoint. Also, for $c_1, c_2 \in G_B(n)$ the balls $B(c'_1, \psi(u_n))$ and $B(c'_2, \psi(u_n))$ with $c'_i \in G_B(n, c_i)$ are disjoint since $B(c'_i, \psi(u_n)) \subset B(c_i, \frac{3}{4}\rho(u_n))$ and $B(c_1, 3\rho(u_n)) \cap B(c_2, 3\rho(u_n)) = \emptyset$. Therefore,

$$m(A_n(\psi, B)) \asymp m(B_n(\psi(u_n))) \; \#G_B(n, c) \; \#G_B(n)$$

and in view of (20) and (24),

$$(25) \qquad m(A_n(\psi, B)) \asymp m(B) \; \times \; \frac{m(B_n(\psi(u_n)))}{m(B_n(\rho(u_n)))} \left(\frac{\rho(u_n)}{\psi(u_n)} \right)^{\gamma} .$$

Finally, let

$$A(\psi, B) := \limsup_{n \to \infty} A_n(\psi, B) := \bigcap_{m=1}^{\infty} \bigcup_{n=m}^{\infty} A_n(\psi, B) .$$

By construction and the fact that ψ is decreasing, we have $A_n(\psi, B) \subset \Delta_l^u(\psi, n) \cap B$ and so $A(\psi, B)$ is a subset of $\Lambda(\psi) \cap B$. Now in view of (18) the proof of Theorem 1 will be completed on showing that

$$(26) \qquad m(A(\psi, B) \cap B) \geq m(B)/C.$$

Notice that estimate (25) on $m(A_n(\psi, B))$ together with the divergent sum hypothesis (6) of the theorem implies that

$$(27) \qquad \sum_{n=1}^{\infty} m(A_n(\psi, B)) = \infty .$$

This is a good sign as if the above sum was to converge, then a simple consequence of the Borel–Cantelli lemma is that $m(A(\psi, B)) = 0$. However, the divergent sum alone is not enough to ensure positive measure; independence of some sort is also required. The following quasi-independence on average will be sufficient.

LEMMA 8 (Quasi–independence on average). *There exists a constant $C > 1$ such that for Q sufficiently large,*

$$\sum_{s,t=1}^{Q} m(A_s(\psi, B) \cap A_t(\psi, B)) \leq \frac{C}{m(B)} \left(\sum_{s=1}^{Q} m(A_s(\psi, B)) \right)^2 .$$

Clearly, Lemma 8 together with the divergent sum (27) and Proposition 2 implies (26). This therefore completes the proof of Theorem 1, assuming of course the quasi–independence on average result which we now prove.

9.2. Proof of Lemma 8 : quasi–independence on average. Throughout, we fix the ball B and write $A_t(\psi)$ for $A_t(\psi, B)$. Also, let $s < t$ and note that

$$
m(A_s(\psi) \cap A_t(\psi)) \;=\; m\left(\bigcup_{c \in G_B(s)} \bigcup_{c' \in G_B(s,c)} B(c', \psi(u_s)) \cap A_t(\psi) \right)
$$

$$
= \sum_{c \in G_B(s)} \sum_{c' \in G_B(s,c)} m\big(B(c', \psi(u_s)) \cap A_t(\psi) \big)
$$

$$
(28) \qquad\qquad \asymp \; \#G_B(s)\, \#G_B(s,c)\, m\big(B_s(\psi(u_s)) \cap A_t(\psi) \big).
$$

We now obtain an upper bound for $m(B_s(\psi(u_s)) \cap A_t(\psi))$ where $B_s(\psi(u_s))$ is by definition any ball of $A_s(\psi)$. Trivially,

$$
m(B_s(\psi(u_s)) \cap A_t(\psi)) \;:=\; m\left(B_s(\psi(u_s)) \cap \bigcup_{c \in G_B(t)} \overset{\circ}{\bigcup_{c' \in G_B(t,c)}} B(c', \psi(u_t)) \right)
$$

$$
(29) \qquad\qquad \asymp \; \sum_{c \in G_B(t)} \sum_{c' \in G_B(t,c)} m\left(B_s(\psi(u_s)) \cap B(c', \psi(u_t)) \right).
$$

We proceed by considering two cases depending on the size of $\psi(u_s)$ compared to $\rho(u_t)$.

Case (i): $t > s$ such that $2\,\psi(u_s) < \rho(u_t)$.   Suppose that there are two elements $c_1,\, c_2 \in G_B(t)$ such that

$$
B_s(\psi(u_s)) \cap B(c_i, \rho(u_t)) \neq \emptyset \qquad (i = \{1, 2\}) \,.
$$

Then, $\mathrm{dist}\,(c_1, c_2) \leq 2\psi(u_s) + 2\rho(u_t) < 3\rho(u_t)$. However, by construction the balls $B(c_i, 3\rho(u_t))$ are disjoint, thus $\mathrm{dist}\,(c_1, c_2) \geq 3\rho(u_t)$. Hence, there is at most one ball $B(c, \rho(u_t))$ with $c \in G_B(t)$ that can possibly intersect $B_s(\psi(u_s))$. Now, (23) together with the upper bound intersection condition implies that

$$
\sum_{c' \in G_B(t,c)} m\left(B_s(\psi(u_s)) \cap B(c', \psi(u_t)) \right) \;\asymp\; m\left(B_s(\psi(u_s)) \;\cap\; \overset{\circ}{\bigcup_{c' \in G_B(t,c)}} B(c', \psi(u_t)) \right)
$$

$$
\ll m\left(B_s(\psi(u_s)) \cap B(c, \tfrac{3}{4}\rho(u_t)) \cap \Delta(R_\alpha, \psi(u_t)) \right)
$$

$$
\ll m(B_t(\psi(u_t))) \left(\frac{\psi(u_s)}{\psi(u_t)} \right)^\gamma .
$$

In view of (29) and the fact that at most one ball $B(c, \rho(u_t))$ with $c \in G_B(t)$ can intersect $B_s(\psi(u_s))$, we have that

$$m\left(B_s(\psi(u_s)) \cap A_t(\psi)\right) \ll m\left(B_t(\psi(u_t))\right) \left(\frac{\psi(u_s)}{\psi(u_t)}\right)^\gamma .$$

This together with (20), (24) and (28) implies that

$$m(A_s(\psi) \cap A_t(\psi)) \ll m(B) \times \frac{m\left(B_t(\psi(u_t))\right)}{m\left(B_s(\rho(u_s))\right)} \left(\frac{\rho(u_s)}{\psi(u_t)}\right)^\gamma .$$

Case (ii): $t > s$ such that $2\,\psi(u_s) \geq \rho(u_t)$.   It follows from (29) that

$$m\left(B_s(\psi(u_s)) \cap A_t(\psi)\right) \ll \sideset{}{^*}\sum_{c' \in G_B(t,c)} \sum m\left(B(c', \psi(u_t))\right)$$

$$\ll m\left(B_t(\psi(u_t))\right) \#G_B(t,c)\, N(t,s) ,$$

where the sum $\sum^*$ is taken over $c \in G_B(t)$ such that $B(c, \rho(u_t)) \cap B_s(\psi(u_s)) \neq \emptyset$ and $N(t,s)$ denotes the number of such c. Clearly, in case (ii) $B(c, \rho(u_t)) \cap B_s(\psi(u_s)) \neq \emptyset$ implies that $B(c, \rho(u_t)) \subset B_s(5\psi(u_s))$. Since the balls $B(c, \rho(u_t))$ with $c \in G_B(t)$ are disjoint we obtain the following trivial estimate

$$N(t,s) \leq \frac{m\left(5B_s(\psi(u_s))\right)}{m\left(B_t(\rho(u_t))\right)} \asymp \frac{m\left(B_s(\psi(u_s))\right)}{m\left(B_t(\rho(u_t))\right)} .$$

Thus

$$m\left(B_s(\psi(u_s)) \cap A_t(\psi)\right) \ll m(B_t(\psi(u_t))) \#G_B(t,c) \frac{m\left(B_s(\psi(u_s))\right)}{m\left(B_t(\rho(u_t))\right)} ,$$

which together with (20), (24), (25) and (28) implies that

$$m(A_s(\psi) \cap A_t(\psi)) \ll \frac{1}{m(B)}\, m(A_s(\psi))\, m(A_t(\psi)).$$

The upshot of these two cases, is that for Q sufficiently large

$$\sum_{s,t=1}^{Q} m(A_s(\psi) \cap A_t(\psi))$$

$$= \sum_{s=1}^{Q} m(A_s(\psi)) + 2 \sum_{s=1}^{Q-1} \sum_{\substack{s+1 \leq t \leq Q \\ \text{case(i)}}} m(A_s(\psi) \cap A_t(\psi))$$

$$+ 2 \sum_{s=1}^{Q-1} \sum_{\substack{s+1 \leq t \leq Q \\ \text{case(ii)}}} m(A_s(\psi) \cap A_t(\psi))$$

$$\ll \sum_{s=1}^{Q} m(A_s(\psi)) + \frac{1}{m(B)} \Big(\sum_{s=1}^{Q} m(A_s(\psi)) \Big)^2$$

$$+ m(B) \sum_{s=1}^{Q-1} \sum_{\substack{s+1 \leq t \leq Q \\ \psi(u_s) < \rho(u_t)}} \frac{m(B_t(\psi(u_t)))}{m(B_s(\rho(u_s)))} \Big(\frac{\rho(u_s)}{\psi(u_t)} \Big)^{\gamma}.$$

By (25) and condition (7) imposed in the statement of the theorem, the latter double sum is $\ll \Big(m(B)^{-1} \sum_{s=1}^{Q} m(A_s(\psi)) \Big)^2$. By (27), for Q sufficiently large $\sum_{s=1}^{Q} m(A_s(\psi)) \leq m(B)^{-1}(\sum_{s=1}^{Q} m(A_s(\psi)))^2$. The statement of Lemma 8 now readily follows and thereby completes the proof of Theorem 1.   #

10. Proof of Theorem 2: $0 \leq G < \infty$

We begin by observing that the case $\gamma = \delta$ is excluded by the various hypotheses imposed on the dimension function f – see also §5. *Thus, without loss of generality we can assume in proving Theorem 2 that*

$$0 \leq \gamma < \delta.$$

To prove Theorem 2 we proceed as follows. For any fixed $\eta \gg 1$ we construct a Cantor subset $\mathbf{K}_\eta$ of $\Lambda(\psi)$ and a probability measure μ supported on $\mathbf{K}_\eta$ satisfying the condition that for an arbitrary ball A of sufficiently small radius $r(A)$

(30) $$\mu(A) \ll \frac{f(r(A))}{\eta},$$

where the implied constant is absolute. By the Mass Distribution Principle, the above inequality implies that $\mathcal{H}^f(\mathbf{K}_\eta) \gg \eta$. Since $\mathbf{K}_\eta \subset \Lambda(\psi)$, we obtain that

$\mathcal{H}^f(\Lambda(\psi)) \gg \eta$. However, $\eta \gg 1$ is arbitrarily large whence $\mathcal{H}^f(\Lambda(\psi)) = \infty$ and this proves Theorem 2.

In view of the above outline, the whole strategy of our proof is centred around the construction of a 'right type' of Cantor set $\mathbf{K}_\eta$ which supports a measure μ with the desired property. The actual nature of the construction of $\mathbf{K}_\eta$ will depend heavily on whether G defined by (9) is finite or infinite. In this section we deal with the case that $0 \le G < \infty$. The case that $G = \infty$ is substantially easier – see §11.

10.1. Preliminaries. In this section we group together for clarity and convenience various concepts and results which will be required in constructing the Cantor set $\mathbf{K}_\eta$. We shall make use of the various hypotheses of Theorem 2 as required. In particular, the measure m is of type (M2).

10.1.1. *The sets* $G_B(n)$ *and* $G_B(n,c)$. Let $B = B(x,r)$ be an arbitrary ball with centre $x \in \Omega$. Assume that its radius r is sufficiently small so that local m-ubiquity and the measure estimate (3) are fulfilled for B. Relabel the sets $G_B(n)$ and $\tilde{G}_\Omega(n)$ constructed in §9 by $G'_B(n)$ and $G'_\Omega(n)$ respectively. By keeping track of constants, the estimate (20) for $\#G'_B(n)$ is explicitly as follows:

$$\frac{a\,\kappa}{b\,(36)^\delta} \left(\frac{r}{\rho(u_n)}\right)^\delta \le \#G'_B(n) \le \frac{b}{a} \left(\frac{r}{\rho(u_n)}\right)^\delta,$$

where a, b and δ are as in (3) and κ is as in (4). Since 'local' implies 'global' ubiquity, the corresponding estimates for $\#G'_\Omega(n)$ are explicitly as follows:

$$\frac{\kappa_1}{b\,9^\delta} \left(\frac{1}{\rho(u_n)}\right)^\delta \le \#G'_\Omega(n) \le \frac{1}{a} \left(\frac{1}{\rho(u_n)}\right)^\delta$$

where $0 < \kappa_1 \le \kappa$ is the global ubiquity constant arising from local ubiquity.

Now let $0 < c_3 := \min\{\frac{a\,\kappa}{b\,(36)^\delta}, \frac{\kappa_1}{b\,9^\delta}\} < 1$ and define $G_B(n)$ to be any sub-collection of $G'_B(n)$ such that

$$\#G_B(n) = \left[c_3 \left(\frac{r}{\rho(u_n)}\right)^\delta\right],$$

where $[x]$ denotes the integer part of a real number x. Thus, for n sufficiently large

$$(31) \qquad \tfrac{1}{2}\,c_3 \left(\frac{r}{\rho(u_n)}\right)^\delta \le \#G_B(n) \le c_3 \left(\frac{r}{\rho(u_n)}\right)^\delta,$$

where we take $r = 1$ when B is replaced by Ω.

In the following the arbitrary ball B can be replaced by the whole space Ω without any loss of generality. Assume for the moment that $\gamma > 0$. Associated with any $c \in G_B(n)$ is the set $G_B(n, c)$ for which (22) and (23) are satisfied. Then the estimate (24) for $\#G_B(n, c)$ is explicitly given by

$$(32) \qquad c_4 \left(\frac{\rho(u_n)}{\psi(u_n)} \right)^\gamma \leq \#G_B(n, c) \leq c_5 \left(\frac{\rho(u_n)}{\psi(u_n)} \right)^\gamma .$$

where $0 < c_4 := \frac{c_1 a}{2^\gamma 9^\delta b} < 1 < c_5 := \frac{c_2 b}{a}$ and c_1 and c_2 are the constants appearing in the intersection conditions. In the case $\gamma = 0$, we define $G_B(n, c) := \{c\}$ and so $\#G_B(n, c) = 1$.

Note that for any distinct $c', c'' \in G_B(n, c)$ we have that $d(c', c'') \geq 3\psi(u_n)$. This follows from the fact that by construction the respective balls of radius $3\psi(u_n)$ are disjoint. Hence, for any $x \in B(c', \psi(u_n))$ and $y \in B(c'', \psi(u_n))$ we have that

$$(33) \qquad d(x, y) \geq 2\psi(u_n) .$$

Also recall that any ball $B(c', \psi(u_n))$ with $c' \in G_B(n, c)$ is contained in $B(c, \frac{3}{4}\rho(u_n))$ and in turn the ball $B(c, \rho(u_n))$ is contained in B.

Remark: In the construction of the set $G_B(n, c)$ we make use of the fact that without loss of generality, $\rho(u_n) > 24\psi(u_n)$ for n sufficiently large. This guarantees that any ball $B(c', 3\psi(u_n))$ with $c' \in G_B(n, c)$ is contained in $B(c, \frac{3}{4}\rho(u_n))$. To see that the above fact remains valid under the hypotheses of Theorem 2 we observe that, without loss of generality, we can assume that $\rho(u_n)^{-1} \psi(u_n) \to 0$ as $n \to \infty$. If this was not the case then $\limsup \rho(u_n)^{-1} \psi(u_n) > 0$ as $n \to \infty$ and since 'local' implies 'global' ubquity, Theorem 1 implies that $m(\Lambda(\psi)) > 0$. In turn this implies that $\mathcal{H}^\delta(\Lambda(\psi))$ is positive and finite. By the elementary fact stated in §7, $\mathcal{H}^f(\Lambda(\psi)) = \infty$ for any dimension function f such that $r^{-\delta} f(r) \to \infty$ as $r \to 0$.

10.1.2. *Working on a subsequence of u.* To begin with recall the following simple facts: (i) if the hypothesis of 'global' or 'local' ubiquity are satisfied for a particular upper sequence u then they are also satisfied for any subsequence s and (ii) if ρ is u-regular then it is s-regular for any subsequence s. Also note that if G is finite, then $\limsup_{n \to \infty} g(s_n) < \infty$ for any subsequence s of u.

Now notice that for any $m \in \mathbb{N}$, if ρ is u-regular with constant $\lambda < 1$ then we can find a subsequence s of u such that

$$\rho(s_{t+1}) \; < \; \lambda^m \, \rho(s_t) \qquad \text{and such that} \qquad \sum g(s_t) = \infty \; .$$

Thus, without loss of generality in establishing part (i) of Theorem 2 ($G = 0$) we can assume that ρ is u-regular with constant λ as small as we please. The existence of such a subsequence s is easy to verify. Trivially, for $p \geq m$ we have that $\rho(u_{n+p}) < \lambda^m \, \rho(u_n)$. For $t \in \mathbb{N}$, let $g(u_{r_t}) := \max\{g(u_r) : m(t-1) < r \leq mt\}$. Then

$$\infty \; = \; \sum_{r=1}^{\infty} g(u_r) \; = \; \sum_{t=1}^{\infty} \sum_{m(t-1)<r\leq mt} g(u_r) \; \leq \; m \sum_{t=1}^{\infty} g(u_{r_t})$$

$$\ll \; \sum_{n=1}^{\infty} g(u_{r_{2n}}) \; + \; \sum_{n=1}^{\infty} g(u_{r_{2n-1}}) \; .$$

Thus on both the sequences $\{u_{r_{2n}}\}$ and $\{u_{r_{2n-1}}\}$ the function ρ satisfies the required regularity condition and for one of them the divergent sum condition is satisfied.

Next notice that if $0 < G < \infty$, then there exists a strictly increasing sequence $\{n_i\}$ such that $g(u_{n_i}) \geq G/2 > 0$. Since $\lim_{r\to\infty} \rho(r) = 0$, it follows that for any $\lambda < 1$ there exists a subsequence s of $\{u_{n_i}\}$ such that $\rho(s_{t+1}) < \lambda \rho(s_t)$ and $\sum g(s_t) = \infty$. Thus, without loss of generality in establishing Theorem 2 for the case that $0 \leq G < \infty$ we can assume that ρ is u regular with constant as small as we please.

10.2. The Cantor set $\mathbf{K}_\eta$. We are assuming that $0 \leq G < \infty$. Let $G^* := \max\{2, \sup_{n\in\mathbb{N}} g(u_n)\}$. Then

$$g(u_n) \; < \; G^* \qquad \text{for all } n \; .$$

Now fix a real number η such that

$$\eta \; > \; G^* \; .$$

To avoid cumbersome expressions, let ϖ denote the following repeatedly occurring constant

$$(34) \qquad\qquad \varpi \; := \; \frac{c_3 c_4 a}{3^\delta \, 32 \, b^2 \, c_2} \; < \; 1 \; .$$

In view of §10.1.2, we can assume that for n sufficiently large $\rho(u_{n+1}) \leq \lambda\rho(u_n)$ with

$$(35) \qquad\qquad 0 \; < \; \lambda \; < \; \left(\frac{a}{a + 3^\delta \, 8 \, b \, c_2} \right)^{\frac{1}{\delta-\gamma}} \; .$$

Finally, unless stated otherwise $B(r)$ will denote a generic ball of radius r centred at a point in Ω.

10.2.1. *Constructing the first level* $\mathbf{K(1)}$. Choose t_1 large enough so that

$$(36) \qquad g(u_{t_1}) \; < \; G^* \; < \; \frac{a}{3^{\delta}\, 8\, c_2\, b}\, \frac{\eta}{\varpi}\, ,$$

$$(37) \qquad \frac{f(\psi(u_{t_1}))}{\psi(u_{t_1})^{\delta}} \; > \; 3^{\delta - \gamma}\, \frac{\eta}{\varpi}\, ,$$

and so that the counting estimate (31) is valid for the set $G_\Omega(t_1)$. Note that the first of these inequalities is possible since $g(u_n) < G^*$ for all n and that $\eta > G^*$. The latter inequality is possible since $f(r)/r^{\delta} \to \infty$ as $r \to 0$. Let $k_1 \geq 1$ be the unique integer such that

$$(38) \qquad \frac{3^{\delta}\, 2\, c_2\, b}{a}\, \frac{\varpi}{\eta}\, \sum_{i=0}^{k_1 - 1} g(u_{t_1 + i}) \; \leq \; \frac{1}{4}$$

$$(39) \qquad \frac{3^{\delta}\, 2\, c_2\, b}{a}\, \frac{\varpi}{\eta}\, \sum_{i=0}^{k_1} g(u_{t_1 + i}) \; > \; \frac{1}{4}$$

Note, the fact that $k_1 \geq 1$ is a consequence of (36).

The first level $\mathbf{K(1)}$ of the Cantor set $\mathbf{K}_\eta$ will now be constructed with the above η in mind. This level will consist of sub-levels $K(t_1 + i)$ where $0 \leq i \leq k_1$.

• **The sub-level** $K(t_1)$ **:**   This consists of balls of common radius $\psi(u_{t_1})$ defined as follows:-

$$K(t_1) \; := \; \bigcup_{c \in G_\Omega(t_1)} \overset{\circ}{\bigcup_{c' \in G_\Omega(t_1, c)}} B(c', \psi(u_{t_1}))\, .$$

• **The sub-level** $K(t_1 + 1)$ **:**   The second sub-level will consist of balls of common radius $\psi(u_{t_1 + 1})$ which substantially avoid balls from the previous sub-level $K(t_1)$. Let

$$h(t_1) \; := \; \left(\frac{\varpi}{\eta}\, \frac{f(\psi(u_{t_1}))}{\psi(u_{t_1})^{\gamma}} \right)^{1/(\delta - \gamma)}\, .$$

Consider some point $c \in G_\Omega(t_1)$. Thus, c lies on a resonant set R_α with $\alpha \in J_l^u(t_1)$. Construct the 'thickening'

$$T_c(t_1) \; := \; \Delta\left(R_\alpha, h(t_1) \right) \cap B(c, \rho(u_{t_1}))\, .$$

Note that in view of (37) we have $3\psi(u_{t_1}) < h(t_1)$, and so by (23)

$$\overset{\circ}{\bigcup_{c' \in G_\Omega(t_1, c)}} B(c', \psi(u_{t_1})) \subset T_c(t_1) \, .$$

Also, notice that in view of (36) we have that $h(t_1) < \rho(u_{t_1})$. Now define

$$T(t_1) := \{T_c(t_1) : c \in G_\Omega(t_1)\} \, ,$$

thus $\#T(t_1) = \#G_\Omega(t_1)$. Clearly, the 'thickenings' in the collection $T(t_1)$ are disjoint since the balls $B(c, 3\rho(u_{t_1}))$ with $c \in G_\Omega(t_1)$ are disjoint. We now introduce a collection of balls from which the next sub-level is to be constructed. Consider the set $G_\Omega(t_1 + 1)$ and for each $c \in G_\Omega(t_1 + 1)$ construct the ball $B(c, \rho(u_{t_1+1}))$. Clearly these balls are disjoint. We disregard any of these balls which lie too close to balls from the previous sub-level. To make this precise, we introduce the sets

$$U_\Omega(t_1 + 1) := \{c \in G_\Omega(t_1 + 1) : B(c, \rho(u_{t_1+1})) \cap T(t_1) \neq \emptyset\}$$

$$V_\Omega(t_1 + 1) := G_\Omega(t_1 + 1) \setminus U_\Omega(t_1 + 1) \, .$$

We claim that $\#V_\Omega(t_1+1) \geq \frac{1}{2} \#G_\Omega(t_1+1)$. This will obviously follow on establishing the upper bound

$$\#U_\Omega(t_1 + 1) < \frac{1}{2} \#G_\Omega(t_1 + 1) \, .$$

There are two cases to consider.

Case (i): $\rho(u_{t_1+1}) < h(t_1)$.   Suppose that $B(c_1, \rho(u_{t_1+1}))$ with $c_1 \in G_\Omega(t_1 + 1)$ intersects some $T_c(t_1) \in T(t_1)$. Then $B(c_1, \rho(u_{t_1+1})) \cap B(c, \rho(u_{t_1})) \neq \emptyset$, and since $\rho(u_{t_1+1}) \leq \rho(u_{t_1})$ we have the inclusion $B(c_1, \rho(u_{t_1+1})) \subset B(c, 3\rho(u_{t_1}))$. Moreover, for $\rho(u_{t_1+1}) < h(t_1)$

$$B(c_1, \rho(u_{t_1+1})) \subset \Delta(R_\alpha, 3h(t_1)) \cap B(c, 3\rho(u_{t_1})) \, .$$

Let N denote the number of balls $B(c_1, \rho(u_{t_1+1}))$ with $c_1 \in G_\Omega(t_1 + 1)$ that can possibly intersect some fixed $T_c(t_1) \in T(t_1)$. Then,

$$m\left(\Delta(R_\alpha, 3h(t_1)) \cap B(c, 3\rho(u_{t_1}))\right) \geq N\, m(B(\rho(u_{t_1+1})))$$
$$\geq N\, a\, \rho(u_{t_1+1})^\delta.$$

For $h(t_1) < \rho(u_{t_1})$, the intersection condition (ii) implies that

$$m\left(\Delta(R_\alpha, 3h(t_1)) \cap B(c, 3\rho(u_{t_1}))\right) \leq c_2\, b\, 3^\delta\, h(t_1)^{\delta-\gamma}\, \rho(u_{t_1})^\gamma \, .$$

Thus,

$$N \leq \frac{c_2 \, b \, 3^\delta}{a} \frac{\varpi}{\eta} \frac{f(\psi(u_{t_1}))}{\psi(u_{t_1})^\gamma} \frac{\rho(u_{t_1})^\gamma}{\rho(u_{t_1+1})^\delta} \, .$$

It follows, by (31) and (36) or equivalently (38), that

$$\#U_\Omega(t_1+1) \leq N \, \#T(t_1) \leq \frac{c_2 \, b \, 3^\delta}{a} \frac{\varpi}{\eta} \frac{f(\psi(u_{t_1}))}{\psi(u_{t_1})^\gamma} \frac{c_3 \, \rho(u_{t_1})^{\gamma-\delta}}{\rho(u_{t_1+1})^\delta}$$

$$\leq \frac{2 \, c_2 \, b \, 3^\delta}{a} \frac{\varpi}{\eta} \, g(u_{t_1}) \, \#G_\Omega(t_1+1) < \frac{1}{4} \#G_\Omega(t_1+1) \, .$$

<u>Case</u> (ii): $\rho(u_{t_1+1}) \geq h(t_1)$.   A similar argument to that given above implies that if $B(c_1, \rho(u_{t_1+1}))$ with $c_1 \in G_\Omega(t_1+1)$ intersects some $T_c(t_1) \in T(t_1)$ then

$$B(c_1, \rho(u_{t_1+1})) \subset \Delta(R_\alpha, 3\rho(u_{t_1+1})) \cap B(c, 3\rho(u_{t_1})) \, .$$

As before, let N denote the number of balls $B(c_1, \rho(u_{t_1+1}))$ with $c_1 \in G_\Omega(t_1+1)$ that can possibly intersect some fixed $T_c(t_1) \in T(t_1)$. Then,

$$m\left(\Delta(R_\alpha, 3\rho(u_{t_1+1})) \cap B(c, 3\rho(u_{t_1}))\right) \geq N \, m(B(\rho(u_{t_1+1})))$$

$$\geq N \, a \, \rho(u_{t_1+1})^\delta.$$

Since $\rho(u_{t_1+1}) < \rho(u_{t_1})$, the intersection condition (ii) implies that

$$m\left(\Delta(R_\alpha, 3\rho(u_{t_1+1})) \cap B(c, 3\rho(u_{t_1}))\right) \leq c_2 \, b \, 3^\delta \, \rho(u_{t_1+1})^{\delta-\gamma} \, \rho(u_{t_1})^\gamma \, .$$

Thus,

$$N \leq \frac{c_2 \, b \, 3^\delta}{a} \left(\frac{\rho(u_{t_1})}{\rho(u_{t_1+1})}\right)^\gamma \, .$$

It follows, by (31) and (35), that

$$\#U_\Omega(t_1+1) \leq N \, \#T(t_1) \leq \frac{c_2 \, b \, 3^\delta}{a} \left(\frac{\rho(u_{t_1})}{\rho(u_{t_1+1})}\right)^\gamma c_3 \left(\frac{1}{\rho(u_{t_1})}\right)^\delta$$

$$\leq \frac{2 \, c_2 \, b \, 3^\delta}{a} \left(\frac{\rho(u_{t_1+1})}{\rho(u_{t_1})}\right)^{\delta-\gamma} \#G_\Omega(t_1+1)$$

$$\leq \frac{2 \, c_2 \, b \, 3^\delta}{a} \lambda^{\delta-\gamma} \#G_\Omega(t_1+1) < \frac{1}{4} \#G_\Omega(t_1+1) \, .$$

On combining the two cases, we have $\#U_\Omega(t_1+1) < \frac{1}{2} \, \#G_\Omega(t_1+1)$. Hence

$$\#V_\Omega(t_1+1) \geq \frac{1}{2} \, \#G_\Omega(t_1+1) \, .$$

The second sub-level is now defined to be:

$$K(t_1 + 1) := \bigcup_{c \in V_\Omega(t_1+1)} \mathring{\bigcup_{c' \in G_\Omega(t_1+1,c)}} B(c', \psi(u_{t_1+1})) \ .$$

Note, by construction $K(t_1) \cap K(t_1 + 1) = \emptyset$.

- **The sub-level $K(t_1 + i)$:**   Suppose $k_1 \geq 2$. Fix $2 \leq i \leq k_1$ and for $1 \leq j \leq i-1$ suppose we have constructed the sub-levels

$$K(t_1 + j) = \bigcup_{c \in V_\Omega(t_1+j)} \mathring{\bigcup_{c' \in G_\Omega(t_1+j,c)}} B(c', \psi(u_{t_1+j})) \ .$$

We proceed to construct $K(t_1 + i)$. For a point $c \in V_\Omega(t_1 + (i-1))$ there exists a resonant set R_α with $\alpha \in J_l^u(t_1 + (i-1))$ such that $c \in R_\alpha$. Let

$$h(t_1 + (i-1)) := \left(\frac{\varpi}{\eta} \frac{f(\psi(u_{t_1+(i-1)}))}{\psi(u_{t_1+(i-1)})^\gamma} \right)^{1/(\delta-\gamma)} ,$$

and construct the 'thickening'

$$T_c(t_1 + (i-1)) := \Delta\left(R_\alpha, h(t_1 + (i-1))\right) \cap B(c, \rho(u_{t_1+(i-1)})) \ .$$

Note that in view of (37) and the fact that $f(r)/r^\delta$ is decreasing

$$3\psi(u_{t_1+(i-1)}) < h(t_1 + (i-1)) ,$$

and so by (23)

$$\mathring{\bigcup_{c' \in G_\Omega(t_1+(i-1),c)}} B(c', \psi(u_{t_1+(i-1)})) \subset T_c(t_1 + (i-1)) \ .$$

Also, notice that in view of (36) and the fact that $g(u_n) < G^*$ for all n we have that $h(t_1 + (i-1)) < \rho(u_{t_1+(i-1)})$. Define

$$T(t_1 + (i-1)) := \{T_c(t_1 + (i-1)) : c \in V_\Omega(t_1 + (i-1))\} \ .$$

Thus, $\#T(t_1 + (i-1)) = \#V_\Omega(t_1 + (i-1)) \leq \#G_\Omega(t_1 + (i-1))$. Clearly the above 'thickenings' in $T(t_1 + (i-1))$ are disjoint since the balls $B(c, 3\rho(u_{t_1+(i-1)}))$ with $c \in G_\Omega(t_1 + (i-1))$ are disjoint. Now introduce the set $G_\Omega(t_1 + i)$ and for each $c \in G_\Omega(t_1 + i)$ construct the ball $B(c, \rho(u_{t_1+i}))$. Obviously these balls are disjoint and we proceed by disregarding any of those which lie too close to balls from any of the previous sub-levels $K(t_1 + j)$. To make this precise, introduce

$$U_\Omega(t_1 + i) := \{c \in G_\Omega(t_1 + i) : B(c, \rho(u_{t_1+i})) \cap \bigcup_{j=0}^{i-1} T(t_1 + j) \neq \emptyset\}$$

$$V_\Omega(t_1 + i) := G_\Omega(t_1 + i) \setminus U_\Omega(t_1 + i) .$$

We claim that $\#V_\Omega(t_1 + i) \geq \frac{1}{2} G_\Omega(t_1 + i)$. This will obviously follow on establishing the upper bound $\#U_\Omega(t_1+i) < \frac{1}{2} G_\Omega(t_1+i)$. As before there are two cases to consider.

<u>Case</u> (i): $0 \leq j \leq i - 1$ such that $\rho(u_{t_1+i}) < h(t_1 + j)$.

<u>Case</u> (ii): $0 \leq j \leq i - 1$ such that $\rho(u_{t_1+i}) \geq h(t_1 + j)$.

On following the arguments as in the $i = 1$ case, we obtain that

$$
\begin{aligned}
\#U_\Omega(t_1 + i) \quad \leq \quad & \sum_{\text{case (i)}} \frac{c_2 \, b \, 3^\delta}{a} \frac{\varpi}{\eta} \frac{f(\psi(u_{t_1+j}))}{\psi(u_{t_1+j})^\gamma} \frac{\rho(u_{t_1+j})^\gamma}{\rho(u_{t_1+i})^\delta} \#T(t_1 + j) \\
& + \sum_{\text{case (ii)}} \frac{c_2 \, b \, 3^\delta}{a} \left(\frac{\rho(u_{t_1+j})}{\rho(u_{t_1+i})} \right)^\gamma \#T(t_1 + j).
\end{aligned}
$$

The contribution from case (i) is:

$$
\begin{aligned}
\leq \quad & \sum_{\text{case (i)}} \frac{c_2 \, b \, 3^\delta}{a} \frac{\varpi}{\eta} \frac{f(\psi(u_{t_1+j}))}{\psi(u_{t_1+j})^\gamma} \frac{\rho(u_{t_1+j})^\gamma}{\rho(u_{t_1+i})^\delta} \#G_\Omega(t_1 + j) \\
\leq \quad & \sum_{j=0}^{k_1-1} \frac{2 \, c_2 \, b \, 3^\delta}{a} \frac{\varpi}{\eta} g(u_{t_1+j}) \#G_\Omega(t_1 + i) \; \leq \; \frac{1}{4} \#G_\Omega(t_1 + i) ,
\end{aligned}
$$

by the choice of k_1 – see (38). The contribution from case (ii) is:

$$\leq \sum_{\text{case (ii)}} \frac{c_2\, b\, 3^\delta}{a} \left(\frac{\rho(u_{t_1+j})}{\rho(u_{t_1+i})}\right)^\gamma \#G_\Omega(t_1+j)$$

$$\leq \frac{2\, c_2\, b\, 3^\delta}{a} \sum_{j=0}^{i-1} \left(\frac{\rho(u_{t_1+i})}{\rho(u_{t_1+j})}\right)^{\delta-\gamma} \#G_\Omega(t_1+i)$$

$$\leq \frac{2\, c_2\, b\, 3^\delta}{a} \sum_{j=0}^{i-1} \lambda^{(i-j)(\delta-\gamma)} \#G_\Omega(t_1+i)$$

$$\leq \frac{2\, c_2\, b\, 3^\delta}{a} \sum_{s=1}^{\infty} (\lambda^{\delta-\gamma})^s \#G_\Omega(t_1+i) \;<\; \frac{1}{4}\, \#G_\Omega(t_1+i)\,,$$

by the choice of λ – see (35). On combining the two cases, we obtain that $\#U_\Omega(t_1+i) < \frac{1}{2}\, \#G_\Omega(t_1+i)$ and so

$$\#V_\Omega(t_1+i) \;\geq\; \frac{1}{2}\, \#G_\Omega(t_1+i)$$

as claimed. The sub-level $K(t_1+i)$ is defined to be:

$$K(t_1+i) \;:=\; \bigcup_{c\in V_\Omega(t_1+i)}\; \overset{\circ}{\bigcup_{c'\in G_\Omega(t_1+1,c)}}\; B(c',\psi(u_{t_1+i}))\,.$$

Also, note that by construction for $0 \leq i \neq j \leq k_1$

$$K(t_1+i) \;\cap\; K(t_1+j) \;=\; \emptyset\,.$$

The first level $\mathbf{K(1)}$ of the Cantor set is defined to be

$$\mathbf{K(1)} \;:=\; \bigcup_{i=0}^{k_1} K(t_1+i)\,.$$

10.2.2. *Constructing the second level* $\mathbf{K(2)}$. The second level of the Cantor set is constructed by 'looking' locally at each ball from the previous level. Thus the second level $\mathbf{K(2)}$ will be defined in terms of local levels $K(2, \mathrm{B})$ associated with B in $\mathbf{K(1)}$.

Choose $t_2 > t_1$ sufficiently large so that for any ball B in $\mathbf{K(1)}$ the counting estimate (31) is valid and so that

$$(40) \qquad\qquad \frac{f(\psi(u_{t_2}))}{\psi(u_{t_2})^\delta} \;>\; 3^{\delta-\gamma}\, \frac{1}{\varpi}\, \frac{f(r(B))}{m(B)}\,.$$

Here and throughout, $r(B)$ denotes the radius of the ball B. Note that for $B \in \mathbf{K}(\mathbf{1})$,

$$\psi(u_{t_1+k_1}) \leq r(B) \leq \psi(u_{t_1}) \ .$$

In view of (37), the fact that $g(u_n) < G^*$ for all n and that $f(r)/r^\delta$ is decreasing as r increases, it is easily verified that

(41) $\qquad g(u_{t_2}) < G^* < \dfrac{a}{3^\delta \, 8 \, c_2 \, b} \dfrac{1}{\varpi} \dfrac{f(r(B))}{m(B)} \qquad \forall \quad B \in \mathbf{K}(\mathbf{1}).$

- **The local sub-level $K(t_2, B)$:** Fix a ball B in $\mathbf{K}(\mathbf{1})$. Thus $B = B(c', \psi(u_{t_1+i}))$ is a ball in the sub-level $K(t_1 + i)$ for some $0 \leq i \leq k_1$. Consider the set $G_B(t_2)$. Each $c \in G_B(t_2)$ gives rise to the set $G_B(t_2, c)$. Let

$$K(t_2, B) := \bigcup_{c \in G_B(t_2)} \overset{\circ}{\bigcup_{c' \in G_B(t_2, c)}} B(c', \psi(u_{t_2})) \ .$$

By construction, $K(t_2, B) \subset B$ and indeed $K(t_2, B)$ defines the first local sub-level associated with B. We now proceed to construct further local sub-levels $K(t_2 + i, B)$ where $1 \leq i \leq k_2(B)$ and $k_2(B)$ is the unique integer such that

(42) $\qquad \dfrac{3^\delta \, 2 \, c_2 \, b \, \varpi}{a} \dfrac{m(B)}{f(r(B))} \displaystyle\sum_{i=0}^{k_2(B)-1} g(u_{t_2+i}) \ \leq \ \dfrac{1}{4}$

(43) $\qquad \dfrac{3^\delta \, 2 \, c_2 \, b \, \varpi}{a} \dfrac{m(B)}{f(r(B))} \displaystyle\sum_{i=0}^{k_2(B)} g(u_{t_2+i}) \ > \ \dfrac{1}{4}$

- **The local sub-level $K(t_2 + 1, B)$:** Consider some point $c \in G_B(t_2)$. Thus c lies on the set $R_\alpha \cap \frac{1}{2}B$ for some $\alpha \in J_l^u(t_2)$. Construct the thickening

$$T_c(t_2, B) := \Delta(R_\alpha, h_B(t_2)) \cap B(c, \rho(u_{t_2})) \ ,$$

where

$$h_B(t_2) := \left(\dfrac{\varpi \, m(B)}{f(r(B))} \dfrac{f(\psi(u_{t_2}))}{\psi(u_{t_2})^\gamma} \right)^{1/(\delta - \gamma)} \ .$$

In view of (40), $3\psi(u_{t_2}) < h_B(t_2)$, and so by (23)

$$\overset{\circ}{\bigcup_{c' \in G_B(t_2,c)}} B(c', \psi(u_{t_2})) \subset T_c(t_2, B) \ .$$

Also, notice that in view of (41) we have that $h_B(t_2) < \rho(u_{t_2})$. Now define

$$T(t_2, B) := \{ T_c(t_2, B) : c \in G_B(t_2) \} \ ,$$

thus $\#T(t_2, B) = \#G_B(t_2)$. Moreover, $K(t_2, B) \subset T(t_2, B)$. Consider the set $G_B(t_2 + 1)$ and for each $c \in G_B(t_2 + 1)$ construct the ball $B(c, \rho(u_{t_2+1}))$. Clearly these balls are disjoint. Introduce the sets

$$U_B(t_2 + 1) \ := \ \{ c \in G_B(t_2 + 1) : B(c, \rho(u_{t_2+1})) \cap T(t_2, B) \neq \emptyset \}$$

$$V_B(t_2 + 1) \ := \ G_B(t_2 + 1) \setminus U_B(t_2 + 1) \ \ .$$

We show that $\#U_B(t_2 + 1) < \frac{1}{2} G_B(t_2 + 1)$ by considering the following two cases:

$\underline{\text{Case (i)}}$: $\rho(u_{t_2+1}) < h_B(t_2)$.

$\underline{\text{Case (ii)}}$: $\rho(u_{t_2+1}) \geq h_B(t_2)$.

As in the construction of the sub-level $K(t_1 + 1)$, we find that for case (i):

$$
\begin{aligned}
\#U_B(t_2 + 1) \ &\leq \ \frac{c_2\,b\,3^\delta\,h_B(t_2)^{\delta-\gamma}}{a}\,\frac{\rho(u_{t_2})^\gamma}{\rho(u_{t_2+1})^\delta}\,\#T(t_2, B) \\
&= \ \frac{c_2\,b\,3^\delta\,\varpi}{a}\,\frac{m(B)}{f(r(B))}\,\frac{f(\psi(u_{t_2}))}{\psi(u_{t_2})^\gamma}\,\frac{\rho(u_{t_2})^\gamma}{\rho(u_{t_2+1})^\delta}\,\#G_B(t_2) \ .
\end{aligned}
$$

By (31) and (41), it follows that

$$
\begin{aligned}
\#U_B(t_2 + 1) \ &\leq \ \frac{2\,c_2\,b\,3^\delta\,\varpi}{a}\,\frac{m(B)}{f(r(B))}\,g(u_{t_2})\,\#G_B(t_2 + 1) \\
&< \ \tfrac{1}{4}\,\#G_B(t_2 + 1) \ .
\end{aligned}
$$

For case (ii) we find that

$$
\begin{aligned}
\#U_B(t_2 + 1) \ &\leq \ \frac{c_2\,b\,3^\delta}{a}\,\left(\frac{\rho(u_{t_2})}{\rho(u_{t_2+1})}\right)^\gamma\,\#T(t_2, B) \\
&= \ \frac{c_2\,b\,3^\delta}{a}\,\left(\frac{\rho(u_{t_2})}{\rho(u_{t_2+1})}\right)^\gamma\,\#G_B(t_2) \ .
\end{aligned}
$$

By (31) and (35), it follows that

$$
\begin{aligned}
\#U_B(t_2 + 1) \quad &\leq \quad \frac{2 \, c_2 \, b \, 3^\delta}{a} \left(\frac{\rho(u_{t_2+1})}{\rho(u_{t_2})} \right)^{\delta - \gamma} \#G_B(t_2 + 1) \\
&\leq \quad \frac{2 \, c_2 \, b \, 3^\delta}{a} \, \lambda^{\delta - \gamma} \, \#G_B(t_2) \quad < \quad \frac{1}{4} \, \#G_B(t_2 + 1) \ .
\end{aligned}
$$

The upshot of these estimates is that

$$
\#V_B(t_2 + 1) \quad \geq \quad \frac{1}{2} \, \#G_B(t_2 + 1) \ .
$$

The second local sub-level associated with B is defined to be

$$
K(t_2 + 1, B) \ := \bigcup_{c \in V_B(t_2+1)} \ \overset{\circ}{\bigcup_{c' \in G_B(t_2+1, \, c)}} \ B(c', \psi(u_{t_2+1})) \ .
$$

Clearly, by construction $K(t_2 + 1, B) \subset B$ and $K(t_2 + 1, B) \cap K(t_2, B) = \emptyset$.

- **The local level $K(2, B)$:** For a fixed ball B in $\mathbf{K}(1)$ continue to construct the local sub-levels $K(t_2 + i, B)$ associated with B for $2 \leq i \leq k_2(B)$, assuming of course that $k_2(B) \geq 2$. Briefly, for $1 \leq j \leq i - 1$ suppose we have already constructed the local sub-levels

$$
K(t_2 + j, B) \ := \bigcup_{c \in V_B(t_2+j)} \ \overset{\circ}{\bigcup_{c' \in G_B(t_2+j, \, c)}} \ B(c', \psi(u_{t_2+j})) \ .
$$

For $c \in V_B(t_2 + (i - 1))$ there exists a resonant set R_α with $\alpha \in J_l^u(t_2 + (i - 1))$ such that c lies on the set $R_\alpha \cap \frac{1}{2}B$. Construct the thickening

$$
T_c(t_2 + (i - 1), B) \ := \ \Delta \left(R_\alpha, h_B(t_2 + (i - 1)) \right) \cap B(c, \rho(u_{t_2+(i-1)}))
$$

where

$$
h_B(t_2 + (i - 1)) \ := \ \left(\frac{\varpi \, m(B)}{f(r(B))} \, \frac{f(\psi(u_{t_2+(i-1)}))}{\psi(u_{t_2+(i-1)})^\gamma} \right)^{1/(\delta - \gamma)} \ ,
$$

and define

$$
T(t_2 + (i - 1), B) \ := \ \{ T_c(t_2 + (i - 1), B) : c \in V_B(t_2 + (i - 1)) \} \ .
$$

Introduce the sets

$$
U_B(t_2 + i) \ := \ \{ c \in G_B(t_2 + i) : B(c, \rho(u_{t_2+i})) \cap \bigcup_{j=0}^{i-1} T(t_2 + j, B) \neq \emptyset \}
$$

$$
V_B(t_2 + i) \ := \ G_B(t_2 + i) \setminus U_B(t_2 + i) \ .
$$

On verifying that $\#V_B(t_2 + i) \geq \frac{1}{2} \#G_B(t_2 + i)$ we define $K(t_2 + i, B)$ in the obvious manner. By construction for $0 \leq i \neq j \leq k_2(B)$ we have that

$$K(t_2 + i, B) \cap K(t_2 + j, B) = \emptyset .$$

The local level $\mathrm{K}(2, \mathrm{B})$ associated with B is defined to be

$$\mathrm{K}(2, \mathrm{B}) := \bigcup_{i=0}^{k_2(B)} K(t_2 + i, B) .$$

In turn, the second level of the Cantor set is defined to be

$$\mathbf{K(2)} := \bigcup_{B \in \mathbf{K(1)}}^{\circ} \mathrm{K}(2, \mathrm{B}) .$$

Clearly, by construction $\mathbf{K(2)} \subset \mathbf{K(1)}$.

10.2.3. *Higher levels* $\mathbf{K(n)}$ *and the Cantor set* $\mathbf{K}_\eta$. Following the procedure for constructing $\mathbf{K(1)}$ and $\mathbf{K(2)}$, for any integer $n \geq 2$ we define the n-th level recursively as follows:

$$\mathbf{K(n)} := \bigcup_{B \in \mathbf{K(n-1)}}^{\circ} \mathrm{K}(\mathrm{n}, \mathrm{B}) ,$$

where

$$\mathrm{K}(\mathrm{n}, \mathrm{B}) := \bigcup_{i=0}^{k_n(B)} K(t_n + i, B)$$

is the n-th local level associated with the ball $B \in \mathbf{K(n-1)}$. Here $t_n > t_{n-1}$ is chosen sufficiently large so that for any ball $B \in \mathbf{K(n-1)}$ the counting estimate (31) is valid and so that

$$(44) \qquad \frac{f(\psi(u_{t_n}))}{\psi(u_{t_n})^\delta} > 3^{\delta - \gamma} \frac{1}{\varpi} \frac{f(r(B))}{m(B)} .$$

Note that $r(B) \leq \psi(u_{t_{n-1}})$ for $B \in \mathbf{K(n-1)}$. In view of (37), the fact that $g(u_n) < G^*$ for all n and that $f(r)/r^\delta$ is decreasing as r increases, we have that

$$(45) \qquad g(u_{t_n}) < G^* < \frac{a}{3^\delta \, 8 \, c_2 \, b} \frac{1}{\varpi} \frac{f(r(B))}{m(B)} \qquad \forall \quad B \in \mathbf{K(n-1)}.$$

Also in the above definition of K(n, B), the quantity $k_n(B)$ is the unique integer ≥ 1 such that

$$(46) \qquad \frac{3^\delta \, 2 \, c_2 \, b \, \varpi}{a} \; \frac{m(B)}{f(r(B))} \sum_{i=0}^{k_n(B)-1} g(u_{t_n+i}) \; \leq \; \frac{1}{4}$$

$$(47) \qquad \frac{3^\delta \, 2 \, c_2 \, b \, \varpi}{a} \; \frac{m(B)}{f(r(B))} \sum_{i=0}^{k_n(B)} g(u_{t_n+i}) \; > \; \frac{1}{4}$$

For completeness and to fix notation, we give a quick sketch of the construction of local sub-levels $K(t_n + i, B)$ associated with $B \in \mathbf{K(n-1)}$. We define the first local sub-level associated with B in the usual manner:

$$K(t_n, B) \; := \; \bigcup_{c \in G_{\mathrm{B}}(t_n)} \overset{\circ}{\bigcup_{c' \in G_{\mathrm{B}}(t_n, c)}} B(c', \psi(u_{t_n})) \; .$$

Any subsequent local sub-level $K(t_n + i, B)$ for $1 \leq i \leq k_n(B)$, is obtained by the following recursive procedure. For $0 \leq j \leq i-1$ suppose we have already constructed the local sub-levels $K(t_n + j, B)$. Consider some point $c \in V_B(t_n + (i-1))$ ($:= G_{\mathrm{B}}(t_n)$ if $i = 1$). Thus c lies on the set $R_\alpha \cap \frac{1}{2}B$ for some $\alpha \in J_l^u(t_n + (i-1))$. Construct the thickening

$$T_c(t_n + (i-1), B) \; := \; \Delta\left(R_\alpha, h_B(t_n + (i-1))\right) \cap B(c, \rho(u_{t_n+(i-1)}))$$

where

$$h_B(t_n + (i-1)) \; := \; \left(\frac{\varpi \, m(B)}{f(r(B))} \; \frac{f(\psi(u_{t_n+(i-1)}))}{\psi(u_{t_n+(i-1)})^\gamma} \right)^{1/(\delta-\gamma)} ,$$

and define

$$T(t_n + (i-1), B) \; := \; \{ T_c(t_n + (i-1), B) : c \in V_B(t_n + (i-1)) \} \; .$$

Then in view of (44) and the fact that $f(r)/r^\delta$ is decreasing

$$(48) \qquad 3 \, \psi(u_{t_n+(i-1)}) \; < \; h_B(t_n + (i-1)) \; ,$$

and so by (23)

$$K(t_n + (i-1), B) \; \subset \; T(t_n + (i-1), B) \; .$$

Also, in view of (45) and that $g(u_n) < G^*$ for all n we have that $h_B(t_n + (i-1)) < \rho(u_{t_n+(i-1)})$. Next, introduce the sets

$$U_B(t_n + i) \;\; := \;\; \{c \in G_B(t_n + i) : B(c, \rho(u_{t_n+i})) \cap \bigcup_{j=0}^{i-1} T(t_n + j, B) \neq \emptyset\}$$

$$V_B(t_n + i) \;\; := \;\; G_B(t_n + i) \setminus U_B(t_n + i) \;\; .$$

As in the previous stages of the construction, it can be verified that $\#V_B(t_n + i) \geq \frac{1}{2}\#G_B(t_n + i)$. Finally, define

$$K(t_n + i, B) \;\; := \;\; \bigcup_{c \in V_B(t_n+i)} \; \overset{\circ}{\bigcup_{c' \in G_B(t_n+i,\, c)}} \; B(c', \psi(u_{t_n+i})) \; .$$

Clearly, for $0 \leq i \neq j \leq k_n(B)$ we have that

$$K(t_n + i, B) \; \cap \; K(t_n + j, B) \; = \; \emptyset \; .$$

Furthermore, by construction the local level $\mathrm{K}(\mathrm{n}, \mathrm{B})$ associated with $B \in \mathbf{K(n-1)}$ is contained in B. Therefore

$$\mathbf{K(n)} \; \subset \; \mathbf{K(n-1)} \; .$$

The Cantor set $\mathbf{K}_\eta$ is simply defined as

$$\mathbf{K}_\eta \; := \; \bigcap_{n=1}^{\infty} \mathbf{K(n)} \; .$$

Trivially

$$\mathbf{K}_\eta \; \subset \; \Lambda(\psi) \; .$$

Before moving on to the construction of a measure μ supported on $\mathbf{K}_\eta$, we prove an important lemma. We adopt the notation that $V_\Omega(t_1) := G_\Omega(t_1)$ and for $n \geq 2$ that $V_B(t_n) := G_B(t_n)$.

LEMMA 9.   (i) *For $0 \leq i \leq k_1$,*

$$\sum_{c \in V_\Omega(t_1+i)} \; \sum_{c' \in G_\Omega(t_1+i,\, c)} f(\psi(u_{t_1+i})) \; \geq \; \frac{c_3\, c_4}{4} \, g(u_{t_1+i})$$

(ii) *For $n \geq 2$, let B be a ball in $\mathbf{K(n-1)}$. Then, for $0 \leq i \leq k_n(B)$*

$$\sum_{c \in V_B(t_n+i)} \; \sum_{c' \in G_B(t_n+i,\, c)} f(\psi(u_{t_n+i})) \; \geq \; \frac{c_3\, c_4}{4\, b} \, g(u_{t_n+i})\, m(B) \; .$$

Proof. For either part we consider $i = 0$ and $i \geq 1$ separately.

(i) For $i = 0$, by (31) and (32)

$$\sum_{c \in V_\Omega(t_1)} \sum_{c' \in G_\Omega(t_1, c)} f(\psi(u_{t_1})) = \#V_\Omega(t_1) \, \#G_\Omega(t_1, c) \, f(\psi(u_{t_1}))$$

$$\geq f(\psi(u_{t_1})) \frac{1}{2} c_3 \left(\frac{1}{\rho(u_{t_1})} \right)^\delta c_4 \left(\frac{\rho(u_{t_1})}{\psi(u_{t_1})} \right)^\gamma = \frac{c_3 \, c_4}{2} g(u_{t_1}).$$

For $1 \leq i \leq k_1$, by (31) and (32) and the fact that $\#V_\Omega(t_1 + i) \geq \frac{1}{2} \#G_\Omega(t_1 + i)$

$$\sum_{c \in V_\Omega(t_1+i)} \sum_{c' \in G_\Omega(t_1+i, c)} f(\psi(u_{t_1+i}))$$

$$\geq f(\psi(u_{t_1+i})) \frac{1}{4} c_3 \left(\frac{1}{\rho(u_{t_1+i})} \right)^\delta c_4 \left(\frac{\rho(u_{t_1+i})}{\psi(u_{t_1+i})} \right)^\gamma = \frac{c_3 \, c_4}{4} g(u_{t_1+i}).$$

(ii) For $i = 0$, by (31) and (32)

$$\sum_{c \in V_B(t_n)} \sum_{c' \in G_B(t_n, c)} f(\psi(u_{t_n}))$$

$$\geq f(\psi(u_{t_n})) \frac{1}{2} c_3 \left(\frac{r(B)}{\rho(u_{t_n})} \right)^\delta c_4 \left(\frac{\rho(u_{t_n})}{\psi(u_{t_n})} \right)^\gamma \geq \frac{c_3 \, c_4}{2 \, b} m(B) \, g(u_{t_n}).$$

For $1 \leq i \leq k_n(B)$, we obtain exactly the same estimate except for an extra '$\frac{1}{2}$' factor since $\#V_B(t_n + i) \geq \frac{1}{2} \#G_B(t_n + i)$.

$$\#$$

10.3. A measure on $\mathbf{K}_\eta$. In this section, we construct a probability measure μ supported on $\mathbf{K}_\eta$ satisfying (30); that is, $\mu(A) \ll f(r(A))/\eta$ for an arbitrary ball A of sufficiently small radius $r(A)$.

Suppose $n \geq 2$ and $B \in \mathbf{K}(\mathbf{n})$. For $1 \leq m \leq n - 1$, let B_m denote the unique ball in $\mathbf{K}(\mathbf{m})$ containing the ball B. With this notation in mind we now define a measure μ. For any $B \in \mathbf{K}(\mathbf{n})$, we attach a weight $\mu(B)$ defined recursively as follows:

For $n = 1$,

$$\mu(B) := \frac{f(r(B))}{\sum_{B' \in \mathbf{K}(\mathbf{1})} f(r(B'))}$$

and for $n \geq 2$,

$$\mu(B) := \frac{f(r(B))}{\sum_{B' \in K(n, B_{n-1})} f(r(B'))} \times \mu(B_{n-1}) \, .$$

This procedure thus defines inductively a mass on any ball appearing in the construction of $\mathbf{K}$. In fact a lot more is true — μ can be further extended to all Borel subsets F of Ω to determine $\mu(F)$ so that μ constructed as above actually defines a measure supported on $\mathbf{K}_\eta$; see Proposition 1.7 [20]. We state this formally as a

Fact. The probability measure μ constructed above is supported on $\mathbf{K}_\eta$ and for any Borel subset F of Ω

$$\mu(F) := \mu(F \cap \mathbf{K}_\eta) = \inf \sum_{B \in \mathcal{B}} \mu(B) \,,$$

where the infimum is taken over all coverings $\mathcal{B}$ of $F \cap \mathbf{K}_\eta$ by balls $B \in \{\mathbf{K}(\mathbf{n}) : n \in \mathbb{N}\}$.

It remains to prove the estimate (30) on the measure μ.

10.3.1. *Measure of a ball in the Cantor construction.* If $B \in \mathbf{K}(\mathbf{n})$ for some $n \in \mathbb{N}$, then by the definition of μ

$$\mu(B) \quad := \quad \frac{f(r(B))}{\sum_{B' \in K(n,B_{n-1})} f(r(B'))} \times \mu(B_{n-1})$$

$$(49) \qquad\qquad = \quad \frac{f(r(B))}{\sum_{B' \in \mathbf{K}(\mathbf{1})} f(r(B'))} \prod_{m=1}^{n-1} \frac{f(r(B_m))}{\sum_{B' \in K(m+1,B_m)} f(r(B'))} \,.$$

The above product term is taken to be one when $n = 1$. To proceed we require the following lemma which gives us a lower bound on the terms in the denominator of the above expression.

LEMMA 10.

$$\sum_{B \in \mathbf{K}(\mathbf{1})} f(r(B)) \geq \eta \quad and \quad \sum_{B \in K(n,B_{n-1})} f(r(B)) \geq f(r(B_{n-1})) \quad (n \geq 2).$$

Proof. By Lemma 9, the choice of k_1 (39) and ϖ (34) it follows that

$$\sum_{B \in \mathbf{K}(\mathbf{1})} f(r(B)) \quad = \quad \sum_{i=0}^{k_1} \sum_{c \in V_\Omega(t_1+i)} \sum_{c' \in G_\Omega(t_1+i,c)} f(\psi(u_{t_1+i}))$$

$$\geq \quad \frac{c_3\,c_4}{4} \sum_{i=0}^{k_1} g(u_{t_1+i}) \quad > \quad \frac{c_3\,c_4\,a}{32\,3^\delta\,c_2\,b} \, \frac{\eta}{\varpi} \quad \geq \eta \,.$$

For $n \geq 2$, by Lemma 9, the choice of $k_n(B_{n-1})$ (47) and ϖ (34) it follows that

$$\sum_{B \in K(n,B_{n-1})} f(r(B)) = \sum_{i=0}^{k_n(B_{n-1})} \sum_{c \in V_{B_{n-1}}(t_n+i)} \sum_{c' \in G_{B_{n-1}}(t_n+i,c)} f(\psi(u_{t_n+i}))$$

$$\geq \frac{c_3 \, c_4}{4 \, b} \, m(B_{n-1}) \sum_{i=0}^{k_n(B_{n-1})} g(u_{t_n+i})$$

$$> \frac{c_3 \, c_4 \, a}{32 \, 3^\delta \, c_2 \, b^2} \, \frac{f(r(B_{n-1}))}{\varpi} \geq f(r(B_{n-1})) . \qquad \#$$

In view of the above lemma, it now follows from (49) that for any ball $B \in \mathbf{K(n)}$

(50) $$\mu(B) \leq \frac{f(r(B))}{\eta} .$$

10.3.2. *Measure of an arbitrary ball A.* We now determine the μ-measure of an arbitrary ball A with radius $r(A) \leq r_o$. The ball A need not be centred at a point of Ω. The aim is to show that:

$$\mu(A) \ll \frac{f(r(A))}{\eta} .$$

The measure μ is supported on $\mathbf{K}_\eta$. Thus, without loss of generality we can assume that $A \cap \mathbf{K}_\eta \neq \emptyset$; otherwise $\mu(A) = 0$ and there is nothing to prove.

We can also assume that for every n large enough A intersects at least two balls in $\mathbf{K(n)}$; since if B is the only ball in $\mathbf{K(n)}$ which has non-empty intersection with A, then in view of (50)

$$\mu(A) \leq \mu(B) \leq \frac{f(r(B))}{\eta} \to 0 \qquad \text{as} \qquad n \to \infty$$

$(r(B) \to 0$ as $n \to \infty)$ and again there is nothing to prove. Thus we may assume that there exists an integer $n \geq 2$ such that A intersects only one ball $\widetilde{B}$ in $\mathbf{K(n-1)}$ and at least two balls from $\mathbf{K(n)}$. The case that A intersects two or more balls from the first level can be excluded by choosing $r(A)$ sufficiently small. This follows from the fact that by construction balls in any one level are disjoint. Furthermore, we can assume that

$$r(A) < r(\widetilde{B}) .$$

Otherwise, since f is increasing $\mu(A) \leq \mu(\widetilde{B}) \leq \frac{f(r(\widetilde{B}))}{\eta} \leq \frac{f(r(A))}{\eta}$ and we are done.

Given that A only intersects the ball $\widetilde{B}$ in $\mathbf{K(n-1)}$, the balls from level $\mathbf{K(n)}$ which intersect A must be contained in the local level

$$\mathrm{K}(\mathrm{n},\widetilde{B}) \; := \; \bigcup_{i=0}^{k_n(\widetilde{B})} K(t_n + i, \widetilde{B}) \; .$$

By construction, any ball $B(c', \psi(u_{t_n+i}))$ from $\mathrm{K}(\mathrm{n},\widetilde{B})$ is contained in some thickening $T_c(t_n + i, \widetilde{B})$. Thus A intersects at least one thickening $T_c(t_n + i, \widetilde{B}) \in T(t_n + i, \widetilde{B})$ for some $0 \leq i \leq k_n(\widetilde{B})$.

Let $K(t_n + i', \widetilde{B})$ be the first local sub-level associated with $\widetilde{B}$ such that

$$K(t_n + i', \widetilde{B}) \cap A \neq \emptyset \; .$$

Of course $0 \leq i \leq k_n(\widetilde{B})$ and by definition, for any $i < i'$

$$K(t_n + i, \widetilde{B}) \cap A = \emptyset \; .$$

Thus, A intersects at least one ball $B(c', \psi(u_{t_n+i'}))$ from $K(t_n + i', \widetilde{B})$ and such balls are indeed the largest balls from the n-th level $\mathbf{K(n)}$ that intersect A. Clearly, A intersects at least one thickening

$$T_* \; := \; T_{c_*}(t_n + i', \widetilde{B})$$

in $T(t_n + i', \widetilde{B})$ or equivalently with $c_* \in V_{\widetilde{B}}(t_n + i')$. We now prove a trivial but crucial geometric lemma.

LEMMA 11. *For $i \geq i'$, if A intersects $B(c', \psi(u_{t_n+i})) \subset T_c(t_n + i, \widetilde{B}) \neq T_*$ then*

$$r(A) \; > \; \tfrac{1}{8}\rho(u_{t_n+i}) \; .$$

Proof. We first establish the lemma for $i > i'$. By definition, $T_c(t_n + i, \widetilde{B}) \subset B(c, \rho(u_{t_n+i}))$ and by construction $B(c, \rho(u_{t_n+i})) \cap T_* = \emptyset$. Also, in view of (23)

$$B(c', \psi(u_{t_n+i})) \subset B(c, \tfrac{3}{4}\rho(u_{t_n+i})) \cap \Delta(R_\alpha, \psi(u_{t_n+i})) \subset T_c(t_n + i, \widetilde{B}).$$

Thus, there exists a point $x \in B(c, \frac{3}{4}\rho(u_{t_n+i})) \cap A$. Let $y \in T_* \cap A$. Since $y \notin B(c, \rho(u_{t_n+i}))$ we have that $d(c, y) > \rho(u_{t_n+i})$. It follows that

$$d(x, y) \ge d(y, c) - d(c, x) > \tfrac{1}{4}\rho(u_{t_n+i}) \ .$$

Since $x, y \in A$, $d(x, y) \le 2\, r(A)$ which together with the above inequality implies that $r(A) > \frac{1}{8}\rho(u_{t_n+i})$. Now suppose $i = i'$. By construction, any thickening $T_c(t_n + i', \widetilde{B}) \subset B(c, \rho(u_{t_n+i'}))$ and the balls $B(c, 3\,\rho(u_{t_n+i'}))$ with $c \in V_{\widetilde{B}}(t_n+i')$ are disjoint. Now let $x \in T_c(t_n + i', \widetilde{B}) \cap A$ and $y \in T_* \cap A$. It is easily verified that $d(x, y) > 2\,\rho(u_{t_n+i'})$ and so $r(A) > \rho(u_{t_n+i'})$.   #

In view of the definition of i' and (50), we have that

$$\mu(A) \ \le\ \sum_{i=i'}^{k_n(\bar{B})} \sum_{c\, \in\, V_{\bar{B}}(t_n+i)} \ \sum_{\substack{c'\, \in\, G_{\bar{B}}(t_n+i,c):\\ B(c',\psi(u_{t_n+i}))\cap A \neq \emptyset}} \mu(B(c', \psi(u_{t_n+i})))$$

$$\text{(51)} \qquad\qquad \le\ \frac{1}{\eta} \sum_{i=i'}^{k_n(\bar{B})} f(\psi(u_{t_n+i})) \sum_{c\, \in\, V_{\bar{B}}(t_n+i)} \ \sum_{\substack{c'\, \in\, G_{\bar{B}}(t_n+i,c):\\ B(c',\psi(u_{t_n+i}))\cap A \neq \emptyset}} 1 \ .$$

To proceed, two separate cases need to be considered:

(i) when A intersects at least two thickenings in $T(t_n + i', \widetilde{B})$

(ii) when A intersects only one thickening in $T(t_n + i', \widetilde{B})$; namely T_*.

<u>Case</u> (i): In view of Lemma 11, if A intersects some $T_c(t_n + i, \widetilde{B}) \in T(t_n + i, \widetilde{B})$ then the ball $B(c, \rho(u_{t_n+i}))$ which contains $T_c(t_n + i, \widetilde{B})$ is itself contained in the ball $17A$. Let N_i denote the number of balls $B(c, \rho(u_{t_n+i}))$ with $c \in V_{\widetilde{B}}(t_n+i)$ that can possibly intersect A. By construction these balls are disjoint – in fact the balls $B(c, 3\,\rho(u_{t_n+i}))$ with $c \in G_\Omega(t_n + i)$ are disjoint. Thus,

$$N_i\, m\,(B(3\,\rho(u_{t_n+i}))) \ \le\ m(17A) \ .$$

Now $A \cap \mathbf{K}_\eta \neq \emptyset$, so there exists a point $x \in \Omega \cap A$. Hence, $17A \subset B(x, 34\, r(A))$ and so $m(17A) \le b\, 34^\delta r(A)^\delta$. the upshot of this is that

$$N_i \ \le\ \frac{b\, 34^\delta}{a} \left(\frac{r(A)}{\rho(u_{t_n+i})}\right)^\delta \ .$$

This implies, via (51) and (32) that

$$
\mu(A) \;\leq\; \frac{1}{\eta} \sum_{i=i'}^{k_n(\tilde{B})} f(\psi(u_{t_n+i})) \sum_{\substack{c \in V_{\tilde{B}}(t_n+i) \\ T_c(t_n+i)\cap A\neq\emptyset}} \#G_{\tilde{B}}(t_n+i,c)
$$

$$
\leq\; \frac{c_5}{\eta} \sum_{i=i'}^{k_n(\tilde{B})} f(\psi(u_{t_n+i})) \left(\frac{\rho(u_{t_n+i})}{\psi(u_{t_n+i})}\right)^{\gamma} N_i
$$

$$
\leq\; \frac{c_5\, b\, 34^{\delta}}{a\,\eta}\, r(A)^{\delta} \sum_{i=0}^{k_n(\tilde{B})} g(u_{t_n+i}) \; .
$$

By (46),

$$
\sum_{i=0}^{k_n(B)-1} g(u_{t_n+i}) \;\leq\; \frac{1}{3^{\delta}\,8\,c_2\,b\,\varpi}\; \frac{f(r(\widetilde{B}))}{r(\widetilde{B})^{\delta}} \;,
$$

and by (45) together with the fact that $g(u_n) < G^*$ for all n

$$
g(u_{t_n+k_n(\tilde{B})}) \;<\; \frac{f(r(\widetilde{B}))}{r(\widetilde{B})^{\delta}}\, \frac{1}{\varpi}\, \frac{1}{3^{\delta}\,8\,c_2\,b} \; .
$$

Hence

$$
(52) \qquad\qquad \mu(A) \;\ll\; \frac{1}{\eta}\, r(A)^{\delta}\, \frac{f(r(\widetilde{B}))}{r(\widetilde{B})^{\delta}} \; .
$$

However, $r(A) < r(\widetilde{B})$ and $f(r)/r^{\delta}$ is decreasing. Thus

$$
\frac{f(r(\widetilde{B}))}{r(\widetilde{B})^{\delta}} \;\leq\; \frac{f(r(A))}{r(A)^{\delta}} \;,
$$

which together with (52) implies the desired inequality; namely (30).

<u>Case (ii)</u>:   By assumption, A only intersects T_* from the collection $T(t_n+i,\widetilde{B})$. On rewriting (51) we have that

$$
\mu(A) \;\leq\; \Sigma_1 := \frac{f(\psi(u_{t_n+i'}))}{\eta} \sum_{\substack{c' \in G_{\tilde{B}}(t_n+i',c_*): \\ B(c',\psi(u_{t_n+i'}))\cap A\neq\emptyset}} 1
$$

$$
(53) \qquad\qquad +\;\; \Sigma_2 := \frac{1}{\eta} \sum_{i=i'}^{k_n(\tilde{B})} f(\psi(u_{t_n+i})) \sum_{c \in V_{\tilde{B}}(t_n+i)} \sum_{\substack{c' \in G_{\tilde{B}}(t_n+i,c): \\ B(c',\psi(u_{t_n+i}))\cap A\neq\emptyset}} 1 \; .
$$

In the case A only intersects T_* the second term $\sum_2$ on the right hand side of (53) is defined to be zero. In any case, we can estimate $\sum_2$ in exactly the same way as in case (i) to obtain that

$$\sum_2 \ll \eta^{-1} f(r(A)) \ .$$

We now deal with the first term $\sum_1$ on the right hand side of (53). First note that if $r(A) \gg \rho(u_{t_n+i'})$ then we are done; since by (32)

$$\sum_1 \le \frac{f(\psi(u_{t_n+i'}))}{\eta} \, \#G_{\widetilde{B}}(t_n + i', c_*) \ \ll \ \frac{1}{\eta} \, g(u_{t_n+i'}) \, \rho(u_{t_n+i'})^\delta$$

$$\ll \ \frac{1}{\eta} \, g(u_{t_n+i'}) \, r(A)^\delta.$$

However, by (45) and the fact that $g(u_n) < G^*$ for all n

$$g(u_{t_n+i'}) \ \ll \ \frac{f(r(\widetilde{B}))}{r(\widetilde{B})^\delta} \ .$$

Hence

$$\sum_1 \ \ll \ \frac{r(A)^\delta}{\eta} \, \frac{f(r(\widetilde{B}))}{r(\widetilde{B})^\delta} \ \le \ \frac{f(r(A))}{\eta} \ ,$$

since $r(A) < r(\widetilde{B})$ and $f(r)/r^\delta$ is decreasing. Thus, without loss of generality we can assume that

(54) $$3\, r(A) \ \le \ \rho(u_{t_n+i'}) \ .$$

Now, A must intersect at least two balls in $K(n, \widetilde{B})$. If at least two of them are contained in T_*, say $B(c', \psi(u_{t_n+i'}))$ and $B(c'', \psi(u_{t_n+i'}))$ with $c' \ne c'' \in G_{\widetilde{B}}(t_n + i', c_*)$, then in view of (33)

$$r(A) \ \ge \ \psi(u_{t_n+i'}) \ .$$

On the other hand, if A intersects only one ball $B(c', \psi(u_{t_n+i'})) \subset T_*$ then it must intersect some other ball $B(c'', \psi(u_{t_n+i})) \subset T_c(t_n+i, \widetilde{B})$ with $i > i'$. By construction,

$$B(c'', \psi(u_{t_n+i})) \ \cap \ T_* \ = \ \emptyset$$

and

$$B(c', \psi(u_{t_n+i'})) \ \subset \ B(c', h_{\widetilde{B}}(t_n + i')) \ \subset \ T_*$$

where $h_{\widetilde{B}}(t_n + i')$ is the 'thickening factor' associated with T_*. Recall that $3\,\psi(u_{t_n+i'}) < h_{\widetilde{B}}(t_n + i')$ – this is (48). Now let $x \in B(c'', \psi(u_{t_n+i})) \cap A$ and $y \in B(c', \psi(u_{t_n+i'})) \cap A$. For $x \notin B(c', h_{\widetilde{B}}(t_n+i'))$ we have that $d(x, c') \ge h_{\widetilde{B}}(t_n+i')$. Thus

$$d(x, y) \ \ge \ d(x, c') - d(c', y) \ > \ 2\,\psi(u_{t_n+i'}) \ ,$$

and so $r(A) > \psi(u_{t_n+i'})$. The upshot of this is that, without loss of generality, we can assume that

(55)
$$r(A) \geq \psi(u_{t_n+i'}) \; .$$

In view of (55), it is easily verified that any ball $B(c', \psi(u_{t_n+i'}))$ which intersects A is in fact contained in the ball $3A$. In particular, in view of (23) any such ball is contained in

$$3A \cap B(c_*, \tfrac{3}{4}\rho(u_{t_n+i'})) \cap \Delta(R_\alpha, \psi(u_{t_n+i'})) \; .$$

Let N denote the number of balls $B(c', \psi(u_{t_n+i'}))$ with $G_{\tilde{B}}(t_n + i', c_*)$ that intersect A. Then

(56)
$$\begin{aligned}
m\left(3A \cap B(c_*, \rho(u_{t_n+i'})) \cap \Delta(R_\alpha, \psi(u_{t_n+i'}))\right) &\geq N\, m\left(B(\psi(u_{t_n+i'}))\right) \\
&\geq a\, N\, \psi(u_{t_n+i'})^\delta \; .
\end{aligned}$$

We would now like to apply the upper bound intersection condition to the quantity on the left hand side of (56). However, this requires the ball A to be centred at a point on a resonant set. The following lemma is required.

LEMMA 12. *Suppose $A \cap \mathbf{K}_\eta \neq \emptyset$. Then there exists a ball $A^* \supseteq A$ with $r(A^*) \leq 3\, r(A)$ and centre a^* on a resonant set.*

Proof. Without loss of generality assume that the centre a of A is not on a resonant set. By construction $\mathbf{K}_\eta$ consists of points arbitrarily close to resonant sets. So if $x \in A \cap \mathbf{K}_\eta$, then there exists some resonant set R_α such that $d(x, R_\alpha) < r(A)$. Thus, $d(a, R_\alpha) \leq d(a, x) + d(x, R_\alpha) < 2\, r(A)$ which implies the existence of some point $a^* \in R_\alpha$ such that $d(a, a^*) < 2\, r(A)$. Now if y is any point of A, then $d(y, a^*) < 3\, r(A)$ and this completes the proof of the lemma.   #

In view of the lemma and (54), the upper bound intersection condition implies that

$$\begin{aligned}
\text{L.H.S. of (56)} &\leq m\left(3A^* \cap B(c_*, \rho(u_{t_n+i'})) \cap \Delta(R_\alpha, \psi(u_{t_n+i'}))\right) \\
&\leq c_2\, b\, 9^\delta\, \psi(u_{t_n+i'})^{\delta-\gamma}\, r(A)^\gamma \; .
\end{aligned}$$

This together with (56) implies that

$$N \le \frac{c_2\, b\, 9^\delta}{a} \left(\frac{r(A)}{\psi(u_{t_n + i'})} \right)^\gamma .$$

Hence, by (55) and the fact that $f(r)/r^\gamma$ is decreasing as $r \to 0$

$$\Sigma_1 := \frac{f(\psi(u_{t_n + i'}))}{\eta}\, N \ll \frac{r(A)^\gamma}{\eta}\, \frac{f(\psi(u_{t_n + i'}))}{\psi(u_{t_n + i'})^\gamma} \ll \frac{f(r(A))}{\eta} .$$

Thus for case (ii),

$$\mu(A) \le \Sigma_1 + \Sigma_2 \ll \frac{f(r(A))}{\eta} .$$

On combining the two cases, we have shown that $\mu(A) \ll f(r(A))/\eta$ for an arbitrary ball A. This completes the proof of Theorem 2 in the case that G is finite.   #

11. Proof of Theorem 2: $G = \infty$

The proof of Theorem 2 in the case that G is infinite follows the same strategy as the proof when G is finite. That is to say, we construct a Cantor subset of $\Lambda(\psi)$ which supports a certain probability measure and then apply the Mass Distribution Theorem. However, to execute this strategy in the case that G is infinite is far simpler than in the finite case. During the proof of the infinite case we shall omit many of the details. After the proof of the finite case in the previous section, the details should pose no real difficulties to the reader.

To start with observe that we can assume, without loss of generality that $0 \le \gamma < \delta$. Also we can assume, without loss of generality that $\lim_{n \to \infty} \psi(u_n)/\rho(u_n) = 0$. In particular, we can assume that $\rho(u_n) > 24\, \psi(u_n)$ for n large enough. Hence, for an arbitrary ball $B = B(x, r)$ with r sufficiently small and $x \in \Omega$ or with $B = \Omega$ we are able to construct the sets $G_B(n)$ and $G_B(n, c)$ as in §10.1.1.

By definition, for each $c \in G_B(n)$ there exists an $\alpha \in J_l^u(n) := \{\alpha \in J : l_n < \beta_\alpha \le u_n\}$ such that $c \in R_\alpha$. In particular, by construction for $c' \in G_B(n, c)$ the ball $B(c', \psi(u_n))$ is contained in the ball $B(c, \frac{3}{4}\rho(u_n))$ and $c' \in R_\alpha$.

11.1. The Cantor set K and the measure μ. We start by defining a Cantor subset **K** of $\Lambda(\psi)$ which is dependent on a certain, strictly increasing sequence of natural numbers $\{t_i : i \in \mathbb{N}\}$.

The Cantor set **K**.   Choose t_1 sufficiently large so that the counting estimate (31) is valid for the set $G_\Omega(t_1)$ and define the first level **K(1)** of the Cantor set **K** as follows:

$$\mathbf{K(1)} \; := \; \bigcup_{c \in G_\Omega(t_1)} \; \overset{\circ}{\bigcup_{c' \in G_\Omega(t_1,c)}} \; B(c', \psi(u_{t_1})) \; .$$

For $n \geq 2$ we define the n-th level **K(n)** recursively as follows:

$$\mathbf{K(n)} \; := \; \overset{\circ}{\bigcup_{B \in \mathbf{K(n-1)}}} \; K(n, B) \; ,$$

where

$$K(n, B) \; := \; \bigcup_{c \in G_B(t_n)} \; \overset{\circ}{\bigcup_{c' \in G_B(t_n,c)}} \; B(c', \psi(u_{t_n}))$$

is the n-th local level associated with the ball $B := B(c', \psi(u_{t_{n-1}})) \in \mathbf{K(n-1)}$. Here $t_n > t_{n-1}$ is chosen sufficiently large so that for any ball B in **K(n − 1)** the counting estimate (31) is valid.

The Cantor set **K** is simply given by

$$\mathbf{K} \; := \; \bigcap_{n=1}^{\infty} \mathbf{K(n)} \; .$$

Trivially,

$$\mathbf{K} \; \subset \; \Lambda(\psi) \; .$$

The measure μ.   Suppose $n \geq 2$ and $B \in \mathbf{K(n)}$. For $1 \leq m \leq n - 1$, let B_m denote the unique ball in **K(m)** containing the ball B. For any $B \in \mathbf{K(n)}$, we attach a weight $\mu(B)$ defined recursively as follows:

For $n = 1$,

$$\mu(B) \; := \; \frac{1}{\#G_\Omega(t_1)} \frac{1}{\#G_\Omega(t_1, c)}$$

and for $n \geq 2$,

$$\mu(B) \; := \; \frac{1}{\#G_{B_{n-1}}(t_n)} \frac{1}{\#G_{B_{n-1}}(t_n, c)} \; \times \; \mu(B_{n-1}) \; .$$

By the definition of μ and the counting estimates (31) and (32), it follows that

$$(57) \qquad \mu(\mathrm{B}) \;\leq\; \rho(u_{t_n})^{\delta-\gamma}\, \psi(u_{t_n})^{\gamma}\, \left(\frac{2}{c_3 c_4}\right)^n \times \prod_{m=1}^{n-1} \left(\frac{\rho(u_{t_m})}{\psi(u_{t_m})}\right)^{\delta-\gamma}$$

$$(58) \qquad\qquad \leq\; f(r(\mathrm{B}))\, c_6^n\, \frac{1}{g(u_{t_n})} \times \prod_{m=1}^{n-1} \left(\frac{\rho(u_{t_m})}{\psi(u_{t_m})}\right)^{\delta-\gamma},$$

where $c_6 := 2/(c_3 c_4) > 1$ is a constant and $r(\mathrm{B}) := \psi(u_{t_n})$. The above product term is taken to be equal to one when $n = 1$.

11.2. Completion of the proof. Fix $\eta \geq 1$. Since $\limsup_{n\to\infty} g(u_n) := G = \infty$, the sequence $\{t_i\}$ associated with the construction of the Cantor set $\mathbf{K}$ can clearly be chosen so that

$$(59) \qquad\qquad \eta \,\times\, c_6^i \,\times\, \prod_{j=1}^{i-1} \left(\frac{\rho(u_{t_j})}{\psi(u_{t_j})}\right)^{\delta-\gamma} \;\leq\; g(u_{t_i}) \;.$$

The product term is one when $i = 1$. It now immediately follows from (58) that for any $\mathrm{B} \in \mathbf{K(n)}$,

$$(60) \qquad\qquad\qquad \mu(\mathrm{B}) \;\leq\; \frac{f(r(\mathrm{B}))}{\eta} \;.$$

We now show that $\mu(A) \ll f(r(A))/\eta$ where A is an arbitrary ball of radius $r(A) \leq r_o$. The same reasoning as in §10.3.2, enables us to assume that $A \cap \mathbf{K} \neq \emptyset$, A is centred on a resonant set, and that there exists an integer $n \geq 2$ such that A intersects only one ball $\widetilde{\mathrm{B}}$ in $\mathbf{K(n-1)}$ and at least two balls from $\mathbf{K(n)}$. Thus, without loss of generality we can assume that

$$(61) \qquad\qquad \psi(u_{t_n}) \;\leq\; r(A) \;\leq\; r(\widetilde{\mathrm{B}}) := \psi(u_{t_{n-1}}) \;.$$

The left hand side of the above inequality makes use of the fact that the balls $3\mathrm{B} := B(c', 3\,\psi(u_{t_n}))$ with $\mathrm{B} \in \mathbf{K(n)}$ are disjoint. Consider the following two cases.

<u>Case</u> (i): $r(A) \leq \rho(u_{t_n})$.   The balls $B(c, 3\,\rho(u_{t_n}))$ with $c \in G_{\widetilde{\mathrm{B}}}(t_n)$ are disjoint. Hence the ball A intersects only one ball $B := B(c, \rho(u_{t_n}))$ with $c \in G_{\widetilde{\mathrm{B}}}(t_n)$. Let N denote the number of balls $B(c', \psi(u_{t_n})) \subset B$ that can possibly intersect A. The upper bound intersection condition implies that

$$m(A \cap 3B \cap \Delta(R_\alpha, \psi(u_{t_n}))) \;\leq\; c_2 \psi(u_{t_n})^{\delta-\gamma}\, r(A)^{\gamma} \;.$$

Thus,

$$N \ \le \ \frac{c_2}{a} \left(\frac{r(A)}{\psi(u_{t_n})} \right)^{\gamma} .$$

In view of (60), (61) and the fact that $f(r)/r^{\gamma}$ is increasing, it follows that

$$\mu(A) \ \le \ N \, \mu(\mathrm{B}) \ \ll \ \frac{1}{\eta} \, r(A)^{\gamma} \, \frac{f(\psi(u_{t_n}))}{\psi(u_{t_n})^{\gamma}} \ \le \ \frac{f(r(A))}{\eta} .$$

<u>Case</u> (ii): $r(A) > \rho(u_{t_n})$.   If A only intersects one ball $B(c, \rho(u_{t_n}))$ with $c \in G_{\widetilde{\mathrm{B}}}(t_n)$, then by (32), (60) and the fact that $f(r)/r^{\gamma}$ is increasing, we have that

$$\mu(A) \ \le \ \mu(B(c, \rho(u_{t_n}))) \ := \ \sum_{c' \in G_{\widetilde{\mathrm{B}}}(t_n, c)} \mu(B(c', \psi(u_{t_n})))$$

$$\le \ c_5 \, \mu(\mathrm{B}) \left(\frac{\rho(u_{t_n})}{\psi(u_{t_n})} \right)^{\gamma} \ll \ \frac{1}{\eta} \, r(A)^{\gamma} \, \frac{f(\psi(u_{t_n}))}{\psi(u_{t_n})^{\gamma}} \ \le \ \frac{f(r(A))}{\eta} .$$

Without loss of generality, assume that A intersects at least two balls $B(c, \rho(u_{t_n}))$ with $c \in G_{\widetilde{\mathrm{B}}}(t_n)$ and let N be the number of such balls that can possibly intersect A. A simple geometric argument, making use of the fact that the balls $B(c, \rho(u_{t_n}))$ are disjoint yields that

$$N \ \le \ \frac{3^{\delta} \, b}{a} \left(\frac{r(A)}{\rho(u_{t_n})} \right)^{\delta} .$$

In view of (32), (57), (59), (61) and the fact that $f(r)/r^{\delta}$ is decreasing, we obtain that

$$\begin{aligned}
\mu(A) \ &\le \ N \, \mu(B(c, \rho(u_{t_n}))) \ = \ N \, \#G_{\widetilde{\mathrm{B}}}(t_n, c) \, \mu(\mathrm{B}) \\
&\ll \ r(A)^{\delta} \left(\frac{2}{c_3 c_4} \right)^{n} \times \prod_{m=1}^{n-1} \left(\frac{\rho(u_{t_m})}{\psi(u_{t_m})} \right)^{\delta - \gamma} \\
&\ll \ f(r(A)) \, c_6^{n} \, \frac{\psi(u_{t_{n-1}})^{\delta}}{f(\psi(u_{t_{n-1}}))} \times \prod_{m=1}^{n-1} \left(\frac{\rho(u_{t_m})}{\psi(u_{t_m})} \right)^{\delta - \gamma} \\
&\ll \ f(r(A)) \, c_6^{n-1} \, \frac{1}{g(u_{t_{n-1}})} \times \prod_{m=1}^{n-2} \left(\frac{\rho(u_{t_m})}{\psi(u_{t_m})} \right)^{\delta - \gamma} \ \le \ \frac{f(r(A))}{\eta} .
\end{aligned}$$

The upshot of these cases is that $\mu(A) \ll f(r(A))/\eta$ for an arbitrary ball A. By the Mass Distribution Principle, $\mathcal{H}^f(\Lambda(\psi)) \ge \mathcal{H}^f(\mathbf{K}) \gg \eta$. However, $\eta \ge 1$ is arbitrary whence $\mathcal{H}^f(\Lambda(\psi)) = \infty$. This thereby completes the proof of Theorem 2 in the case that G is infinite.

$$\#$$

12. Applications

Unless stated otherwise, in all the following statements the convergent parts are easy to establish – just use the 'natural cover' given by the lim sup set under consideration. Also we will make use of the following simple fact. Suppose that $h : \mathbb{R}^+ \to \mathbb{R}^+$ is a real, positive monotonic function, $\alpha \in \mathbb{R}$ and $k > 1$. Then the divergence and convergence properties of the sums

$$\sum_{n=1}^{\infty} k^{n \alpha} \, h(k^n) \qquad \text{and} \qquad \sum_{r=1}^{\infty} r^{\alpha - 1} \, h(r) \qquad \text{coincide.}$$

The various applications have been chosen to illustrate the versatility of our general framework. There are many other applications, such as to inhomogeneous Diophantine approximation, Markov maps and iterated function schemes, which we have decided not to include – mainly to avoid repetition and to keep the length of the paper manageable. Throughout, 'i.m.' is short for 'infinitely many'.

12.1. Linear Forms . Let ψ be an approximating function. An $m \times n$ matrix $X = (x_{ij}) \in \mathbf{I}^{mn} := [0,1]^{mn}$ is said to be ψ–well approximable if the system of inequalities

$$|q_1 \, x_{1j} + q_2 \, x_{2j} + \cdots + q_m \, x_{mj} \ - \ p_j| \ < \ \psi(|\mathbf{q}|) \, |\mathbf{q}| \qquad (1 \le j \le n)$$

is satisfied for infinitely many vectors $\mathbf{q} \in \mathbb{Z}^m$, $\mathbf{p} \in \mathbb{Z}^n$. Here $|\mathbf{q}|$ denotes the supremum norm of the vector $\mathbf{q}$; i.e. $|\mathbf{q}| = \max\{|q_1|, \ldots, |q_m|\}$. The system

$$q_1 \, x_{1j} + q_2 \, x_{2j} + \cdots + q_m \, x_{mj} \qquad (1 \le j \le n)$$

of n real linear forms in m variables $q_1, \ldots, q_m$ will be written more concisely as $\mathbf{q} X$, where the matrix X is regarded as a point in $\mathbf{I}^{mn}$. In view of this notation, the set of ψ–well approximable points will be denoted by

$$W(m, n; \psi) := \{X \in \mathbf{I}^{mn} : |\mathbf{q} X - \mathbf{p}| < \psi(|\mathbf{q}|) \, |\mathbf{q}| \text{ for i.m. } (\mathbf{p}, \mathbf{q}) \in \mathbb{Z}^n \times \mathbb{Z}^m\} .$$

By definition, $|\mathbf{q} X - \mathbf{p}| = \max_{1 \le j \le n} |\mathbf{q}.X^{(j)} - p_j|$ where $X^{(j)}$ is the j'th column vector of X. Note that when $m = n = 1$, the set $W(1, 1; \psi)$ corresponds to the classical set $W(\psi)$ in the one dimensional theory.

With reference to our general framework, let $\Omega := \mathbf{I}^{mn}$, $J := \{(\mathbf{p}, \mathbf{q}) \in \mathbb{Z}^n \times \mathbb{Z}^m \setminus \{\mathbf{0}\} : |\mathbf{p}| \le |\mathbf{q}|\}$, $\alpha := (\mathbf{p}, \mathbf{q}) \in J$, $\beta_\alpha := |\mathbf{q}|$ and $R_\alpha := \{X \in \mathbf{I}^{mn} : \mathbf{q} X = \mathbf{p}\}$.

Thus, the family $\mathcal{R}$ of resonant sets R_α consists of $(m-1)n$–dimensional, rational hyperplanes. Furthermore, $\Delta(R_\alpha, \psi(\beta_\alpha)) := \{X \in \mathbf{I}^{mn} : \mathrm{dist}(X, R_\alpha) < \psi(|\mathbf{q}|)\}$ and

$$\Delta_l^u(\psi, n) := \bigcup_{2^{n-1} < |\mathbf{q}| \leq 2^n} \bigcup_{|\mathbf{p}| \leq |\mathbf{q}|} \Delta(R_\alpha, \psi(\beta_\alpha)) \ .$$

Then

$$W(m, n; \psi) = \Lambda(\psi) := \limsup_{n \to \infty} \Delta_l^u(\psi, n) \ .$$

Now let the measure m be mn–dimensional Lebesgue measure, $\delta = mn$ and $\gamma = (m-1)n$. Then a probabilistic argument involving mean and variance techniques yields the following statement [16, §4.3].

PROPOSITION 4. *The pair $(\mathcal{R}, \beta)$ is a local m–ubiquitous system relative to (ρ, l, u), where $l_{t+1} = u_t := 2^t$ $(t \in \mathbb{N})$ and $\rho : r \to \rho(r) = \text{constant} \times r^{-(m+n)/n}\omega(r)$. Here ω is any real, positive increasing function such that $1/\omega(r) \to 0$ as $r \to \infty$ and such that for any $C > 1$ and r sufficiently large $\omega(2r) < C\omega(r)$.*

In view of the proposition and the fact that the measure m is of type (M2) and that ρ is u-regular, Corollary 2 of Theorem 1 and Theorem 2 yield the divergent parts of the following statements.

THEOREM (KHINTCHINE–GROSHEV). *Let ψ be a real, positive decreasing function. Then*

$$m(W(m, n; \psi)) = \begin{cases} 0 & \text{if} \quad \sum_{r=1}^{\infty} \psi(r)^n \ r^{m+n-1} < \infty \ , \\[2mm] 1 & \text{if} \quad \sum_{r=1}^{\infty} \psi(r)^n \ r^{m+n-1} = \infty \ . \end{cases}$$

THEOREM DV (1997). *Let f be a dimension function such that $r^{-mn} f(r) \to \infty$ as $r \to 0$ and $r^{-mn} f(r)$ is decreasing. Furthermore, suppose that $r^{-(m-1)n} f(r)$ is increasing. Let ψ be a real, positive decreasing function. Then*

$$\mathcal{H}^f(W(m, n; \psi)) = \begin{cases} 0 & \text{if} \quad \sum f(\psi(r)) \psi(r)^{-(m-1)n} \ r^{m+n-1} < \infty, \\[2mm] \infty & \text{if} \quad \sum f(\psi(r)) \psi(r)^{-(m-1)n} \ r^{m+n-1} = \infty. \end{cases}$$

Notice that the function ω associated with ρ does not appear in the above statements. This is a consequence of choosing ω appropriately. With ρ as in the proposition, Corollary 2 implies that

$$m(W(m,n;\psi)) = 1 \quad \text{if} \quad \sum_\omega := \sum \psi(r)^n \ r^{m+n-1} \ \omega(r)^{-n} = \infty \ .$$

However, to obtain the precise statement of the Khintchine-Groshev theorem we need to remove the ω factor. To do this we choose ω in such a way that the divergence/convergence properties of $\sum_\omega$ and $\sum := \sum \psi(r)^n \ r^{m+n-1}$ are the same. It is always possible to find such a function. Clearly, if $\sum_\omega = \infty$ then $\sum = \infty$. On the other hand, if $\sum = \infty$, then we can find a strictly increasing sequence of positive integers $\{r_i\}_{i\in\mathbb{N}}$ such that

$$\sum_{r_{i-1} < r \leq r_i} \psi(r)^n \ r^{m+n-1} > 1 \ ,$$

and $r_i > 2r_{i-1}$. Now simply define ω be the step function $\omega(r) := i^{\frac{1}{n}}$ for $r_{i-1} < r \leq r_i$ and this satisfies the required properties. A similar argument allows us to conclude Theorem DV from Theorem 2 without the presence of the ω factor.

Remark. The above theorems remain valid if the set $W(m,n;\psi)$ is replaced by its 'inhomogeneous' analogue. Briefly, for a vector $\mathbf{b} \in \mathbf{I}^n := [0,1]^n$ consider the set $W_\mathbf{b}(m,n;\psi) := \{X \in \mathbf{I}^{mn} : |\mathbf{q}X - \mathbf{p} - \mathbf{b}| < \psi(|\mathbf{q}|)\,|\mathbf{q}| \text{ for i.m. } (\mathbf{p},\mathbf{q}) \in \mathbb{Z}^n \times \mathbb{Z}^m\}$. Obviously, the 'homogeneous' set $W(m,n;\psi)$ corresponds to the case when $\mathbf{b}$ is the zero vector. Now define $\Lambda(\psi)$ as above with the only modification being that the family $\mathcal{R}$ of resonant sets R_α now consists of $(m-1)n$-dimensional hyperplanes of the form $\{X \in \mathbf{I}^{mn} : \mathbf{q}X = \mathbf{p} + \mathbf{b}\}$. Then, it is possible to show that Proposition 4 remains valid for the pair $(\mathcal{R}, \beta)$ and so Corollary 2 of Theorem 1 and Theorem 2 yield the divergent parts of the Khintchine-Groshev theorem and Theorem DV for the set $W_\mathbf{b}(m,n;\psi)$. The proof of the proposition in the inhomogeneous setup follows from Theorem 2 in [13] if $m = 1$ and from the mean-variance argument carried out in §3.2 of [32] if $m \geq 2$.

12.2. Algebraic Numbers. Let $H(a)$ denote the height of an algebraic number a, that is the maximum of the absolute values of the relatively prime integer coefficients in its minimal defining polynomial. For $d \in \mathbb{N}$, denote by $A(d)$ the set of algebraic numbers a with degree at most n. Given an approximating function ψ, let

$$K_d(\psi) := \{\xi \in [0,1] : |\xi - a| < \psi(H(a)) \text{ for i.m. } a \in A(d)\} \ .$$

The set $K_d(\psi)$ is a generalization of $W(\psi)$ since the rationals are algebraic with degree one. In the case that $\psi(r) = r^{-(d+1)\tau}$ let us write $K_d(\tau)$ for $K_d(\psi)$. A. Baker and W.M. Schmidt [2] have obtained the following analogue of the classical Jarník–Besicovitch Theorem.

BAKER–SCHMIDT THEOREM (1970) .   *For $\tau \geq 1$, $\dim K_d(\tau) = 1/\tau$.*

As an application of our main theorems we are able to give a complete measure theoretic description of $K_d(\psi)$ which not only implies the Baker-Schmidt Theorem but also shows that $\mathcal{H}^{1/\tau}(K_d(\tau)) = \infty$.

Let $\Omega := [0,1]$, $J := \{a : a \in A(d)\}$, $\alpha := a \in J$, $\beta_\alpha := H(a)$ and $R_\alpha := a$. Thus, the family $\mathcal{R}$ of resonant sets R_α consists of points corresponding to algebraic numbers $a \in A(d)$. Furthermore, $\Delta(R_\alpha, \psi(\beta_\alpha)) := B(a, \psi(H(a)))$ and

$$\Delta_l^u(\psi, n) := \bigcup_{a \in J_l^u(n)} B\left(a, \psi(H(a))\right) ,$$

where $J_l^u(n) := \{a \in A(d) : k^{n-1} < H(a) \leq k^n\}$. Here $k > 1$ is a constant. Then

$$K_d(\psi) = \Lambda(\psi) := \limsup_{n \to \infty} \Delta_l^u(\psi, n) .$$

Now let m be one–dimensional Lebesgue measure, $\delta = 1$ and $\gamma = 0$.

PROPOSITION 5. *The pair $(\mathcal{R}, \beta)$ is a local m–ubiquitous system relative to (ρ, l, u) where for $k > k_0$ – a positive absolute constant, $l_{t+1} = u_t := k^t$ $(t \in \mathbb{N})$ and $\rho : r \to \rho(r) := \text{constant} \times r^{-(d+1)}$.*

Baker and Schmidt [2] established Proposition 5 with $\rho(r) = r^{-(d+1)} \times (\log r)^{3d(d+1)}$. This is sufficient only to determine the dimension result (simply apply Corollary 5). The presence of the log term in their ubiquity function ρ rules out the possibility of obtaining the more desirable measure theoretic laws for $K_d(\psi)$. However, a more subtle analysis enables one to remove the log term [3].

In view of the proposition and the fact that the measure m is of type (M2) and that ρ is u-regular, Corollary 2 of Theorem 1 and Theorem 2 yield the divergent parts of the following statements.

THEOREM 3. *Let ψ be a real, positive decreasing function. Then*

$$m\left(K_d(\psi)\right) = \begin{cases} 0 & \text{if} \quad \sum_{r=1}^{\infty} \psi\left(r\right) \ r^d < \infty \ , \\ \\ 1 & \text{if} \quad \sum_{r=1}^{\infty} \psi\left(r\right) \ r^d = \infty \ . \end{cases}$$

THEOREM 4. *Let f be a dimension function such that $r^{-1} f(r) \to \infty$ as $r \to 0$ and $r^{-1} f(r)$ is decreasing. Let ψ be a real, positive decreasing function. Then*

$$\mathcal{H}^f\left(K_d(\psi))\right) = \begin{cases} 0 & \text{if} \quad \sum_{r=1}^{\infty} f\left(\psi(r)\right) \ r^d < \infty \ , \\ \\ \infty & \text{if} \quad \sum_{r=1}^{\infty} f\left(\psi(r)\right) \ r^d = \infty \ . \end{cases}$$

Theorem 3 was first established in [**3**]. As mentioned above, Theorem 4 not only implies the Baker-Schmidt Theorem but also shows that $\mathcal{H}^{1/\tau}\left(K_d(\tau)\right) = \infty$. A weaker form of Theorem 4, has been recently established in [**12**].

12.3. Kleinian Groups . The classical results of Diophantine approximation, in particular those from the one dimensional theory, have natural counterparts and extensions in the hyperbolic space setting. In this setting, instead of approximating real numbers by rationals, one approximates limit points of a fixed Kleinian group G by points in the orbit (under the group) of a certain distinguished limit point y. Beardon and Maskit have shown that the geometry of the group is reflected in the approximation properties of points in the limit set. The elements of G are orientation preserving Möbius transformations of the $(n+1)$–dimensional unit ball B^{n+1}. Let Λ denote the limit set of G and let δ denote the Hausdorff dimension of Λ. For any element g in G we shall use the notation $L_g := |g'(0)|^{-1}$, where $|g'(0)|$ is the (Euclidean) conformal dilation of g at the origin.

Let ψ be an approximating function and let

$$W_y(\psi) := \{\xi \in \Lambda : |\xi - g(y)| < \psi(L_g) \text{ for i.m. } g \text{ in } G\}.$$

This is the set of points in the limit set Λ which are 'very close' to infinitely many images of a 'distinguished' point y. The 'closeness' is of course governed by the approximating function ψ. The limit point y is taken to be a parabolic fixed point if the group has parabolic elements and a hyperbolic fixed point otherwise.

Geometrically finite groups: Let us assume that the geometrically finite group has parabolic elements so it is not convex co-compact. Thus our distinguished limit point y is a parabolic fixed point, say p. Associated with p is a geometrically motivated set $\mathcal{T}_p$ of coset representatives of $G_p \backslash G := \{gG_p : g \in G\}$; so chosen that if $g \in \mathcal{T}_p$ then the orbit point $g(0)$ of the origin lies within a bounded hyperbolic distance from the top of the standard horoball $H_{g(p)}$. The latter, is an $(n+1)$–dimensional Euclidean ball contained in B^{n+1} such that its boundary touches the unit ball S^n at the point $g(p)$. Let R_g denote the Euclidean radius of $H_{g(p)}$. As a consequence of the definition of $\mathcal{T}_p$, it follows that

$$\frac{1}{C\,L_g} \leq R_g \leq \frac{C}{L_g}$$

where $C > 1$ is an absolute constant. Also, it is worth mentioning that the balls in the standard set of horoballs $\{H_{g(p)} : g \in \mathcal{T}_p\}$ corresponding to the parabolic fixed point p are pairwise disjoint. For further details and references regarding the above notions and statements see any of the papers [**26, 35, 40**]. With reference to our general framework, let $\Omega := \Lambda$, $J := \{g : g \in \mathcal{T}_p\}$, $\alpha := g \in J$, $\beta_\alpha := C\,R_g^{-1}$ and $R_\alpha := g(p)$. Thus, the family $\mathcal{R}$ of resonant sets R_α consists of orbit points $g(p)$ with $g \in \mathcal{T}_p$. Furthermore, $\Delta(R_\alpha, \psi(\beta_\alpha)) := B(g(p), \psi(C\,R_g^{-1}))$ and

$$\Delta_l^u(\psi, n) := \bigcup_{g \in J_l^u(n)} B\left(g(p), \psi(C\,R_g^{-1})\right) ,$$

where $J_l^u(n) := \{g \in \mathcal{T}_p : k^{n-1} < C\,R_g^{-1} \leq k^n\}$. Here $k > 1$ is a constant. Then

$$W_p(\psi) \supset \Lambda(\psi) := \limsup_{n \to \infty} \Delta_l^u(\psi, n) .$$

Now, let m be Patterson measure, $\delta = \dim \Lambda$ and $\gamma = 0$. Thus m is a non-atomic, δ–conformal probability measure supported on Λ. Furthermore, m is of type (M1) with respect to the sequences $l := \{k^{t-1}\}$ and $u := \{k^t\}$ for any $k > 1$ – see below. In fact, the condition that $m(B(c, 2r)) \ll m(B(c, r))$ for balls centred at resonant points is valid for any $c \in \Lambda$. We have the following statement concerning local ubiquity.

PROPOSITION 6. *The pair $(\mathcal{R}, \beta)$ is a local m–ubiquitous system relative to (ρ, l, u) where for $k \geq k_o$ – a positive group constant,*

$$l_{t+1} = u_t := k^t \ \ (t \in \mathbb{N}) \quad and \quad \rho : r \to \rho(r) := \text{constant} \times r^{-1} .$$

The proposition follows from the following three facts which can be found in [**26**, **35**].

- *Local Horoball Counting Result:* Let B be an arbitrary Euclidean ball in S^n centred at a limit point. For $\lambda \in (0,1)$ and $r \in \mathbb{R}^+$ define

$$A_\lambda(B,R) := \{g \in \mathcal{T}_p : g(p) \in B \text{ and } \lambda R \le R_g < R\} \, .$$

There exists a positive group constant λ_o such that if $\lambda \le \lambda_o$ and $R < R_o(B)$, then

$$k_1^{-1} R^{-\delta} m(B) \ \le \ \#A_\lambda(B,R) \ \le \ k_1 R^{-\delta} m(B) \, ,$$

where k_1 is a positive constant independent of B and $R_o(B)$ is a sufficiently small positive constant which does depend on B.

- *Disjointness Lemma:* For distinct elements $g, h \in \mathcal{T}_p$ with $\lambda < R_g/R_h < \lambda^{-1}$, one has $B(g(p), \lambda R_g) \cap B(h(p), \lambda R_h) = \emptyset$.

- *Measure of balls centred at parabolic points:* For $g \in \mathcal{T}_p$ and $r \le R_g$

$$k_2^{-1} r^{2\delta - \mathrm{rk}(p)} R_g^{\mathrm{rk}(p)-\delta} \ \le \ m\left(B(g(p),r)\right) \ \le \ k_2 r^{2\delta-\mathrm{rk}(p)} R_g^{\mathrm{rk}(p)-\delta} \, ,$$

where $\mathrm{rk}(p)$ denotes the rank of the parabolic fixed point p and $k_2 > 1$ is a positive constant independent of g and r. Clearly, this implies that m satisfies condition (M1) with respect to the sequences l and u.

To prove the proposition, let $\rho(r) := C(k\,r)^{-1}$ where $k := 1/\lambda > 1/\lambda_o$ and B be an arbitrary ball centred at a limit point. Then for n sufficiently large

$$m\Big(\, B \cap \overset{\circ}{\bigcup_{\substack{g \in \mathcal{T}_p: \\ k^{n-1} < C\, R_g^{-1} \le k^n}}} B\left(g(p), \rho(k^n)\right)\Big) \ \ge \ m\Big(\, \overset{\circ}{\bigcup_{\substack{g \in \mathcal{T}_p:\, g(p) \in \frac{1}{2}B \\ k^{n-1} < C\, R_g^{-1} \le k^n}}} B\left(g(p), \rho(k^n)\right)\Big)$$

$$\gg \ k^{-n\,\delta} \ \#A_{\frac{1}{k}}\left(\tfrac{1}{2}B\, , C\, k^{-(n-1)}\right)$$

$$\gg \ m(\tfrac{1}{2}B) \ \gg \ m(B) \, .$$

Now let ψ be an approximating function and assume without loss of generality that $\psi(k^n) \le \rho(k^n)$ for n sufficiently large. If this were not the case then the $\limsup$ condition of Theorem 1 can be invoked to imply the desired result below. Since

$\psi(k^n) < R_g$ for $g \in J^u_l(n)$, the above measure fact for balls centred at parabolic points implies that for any $g \in J^u_l(n)$

$$m\left(B\left(g(p), \psi(k^n)\right)\right) \asymp \psi(k^n)^{2\delta - \mathrm{rk}(p)} \, k^{-n(\mathrm{rk}(p) - \delta)} \, .$$

Also notice that $m\left(B\left(g(p), \rho(k^n)\right)\right) \asymp k^{-n\delta}$. It therefore follows that

$$
\begin{aligned}
\text{L.H.S. of (7)} \quad &\leq \quad \sum_{s=1}^{Q-1} k^{s\delta} \sum_{s+1 \leq t \leq Q} \psi(k^t)^{2\delta - \mathrm{rk}(p)} \, k^{-t(\mathrm{rk}(p) - \delta)} \\
&= \quad \sum_{m=2}^{Q} \psi(k^m)^{2\delta - \mathrm{rk}(p)} \, k^{-m(\mathrm{rk}(p) - \delta)} \sum_{r=1}^{m-1} k^{r\delta} \\
&\ll \quad \sum_{m=2}^{Q} \psi(k^m)^{2\delta - \mathrm{rk}(p)} \, k^{m(2\delta - \mathrm{rk}(p))} \quad \ll \quad \text{R.H.S. of (7)} \, .
\end{aligned}
$$

Thus, in view of Proposition 6 and the fact that the measure m is of type (M1) and that ρ is u-regular, Theorem 1 yields the divergent part of the following statement.

THEOREM 5. *Let G be a geometrically finite Kleinian group with parabolic elements and let $\mathrm{rk}(p)$ denote the rank of the parabolic fixed point p. Let ψ be a real, positive decreasing function. Then*

$$
m(W_p(\psi)) = \begin{cases} 0 & \text{if} \quad \sum_{r=1}^{\infty} \psi\left(r\right)^{2\delta - \mathrm{rk}(p)} \, r^{2\delta - \mathrm{rk}(p) - 1} < \infty \, , \\[2ex] 1 & \text{if} \quad \sum_{r=1}^{\infty} \psi\left(r\right)^{2\delta - \mathrm{rk}(p)} \, r^{2\delta - \mathrm{rk}(p) - 1} = \infty \, . \end{cases}
$$

Theorem 5 is not new. However, in previous statements of the theorem a certain regularity condition on ψ is assumed [36, 40, 41]. In Theorem 5, the regularity is removed and replaced by the 'natural' condition that ψ is decreasing. Thus the above Khintchine type theorem is the perfect analogue of the classical statement.

In general, for geometrically finite Kleinian groups with parabolic elements, Patterson measure m is not of type (M2). Thus, Theorem 2 is not applicable even though we have local m-ubiquity (Proposition 6). In fact, in general m is not even of type (M2′) so Theorem 2′ of §5 is not applicable either. However, if the group is of the first kind (so $\Lambda = S^n$) then m is normalized n-dimensional Lebesgue measure on the unit sphere S^n and so is certainly of type (M2). Also, for groups of the first kind $\delta = n = \mathrm{rk}(p)$. Thus, for such groups Theorem 2 yields the divergent part of the following statement.

THEOREM 6. *Let G be a geometrically finite Kleinian group of the first kind with parabolic elements and p be a parabolic fixed point. Let f be a dimension function such that $r^{-n} f(r) \to \infty$ as $r \to 0$ and $r^{-n} f(r)$ is decreasing. Let ψ be a real, positive decreasing function. Then*

$$\mathcal{H}^f \left(W_p(\psi) \right) = \begin{cases} 0 & \text{if} \quad \sum_{r=1}^{\infty} f \left(\psi(r) \right) \; r^{n-1} < \infty , \\ \\ \infty & \text{if} \quad \sum_{r=1}^{\infty} f \left(\psi(r) \right) \; r^{n-1} = \infty . \end{cases}$$

Regarding Theorem 6, all that was previously known were dimension statements for $W_p(\psi)$ [**43**]. For example, in the case $\psi(r) = r^{-\tau}$ let us write $W_p(\tau)$ for $W_p(\psi)$. Then $\dim W_p(\tau) = n/\tau$ $(\tau \geq 1)$. Clearly, Theorem 6 implies this statement and shows that the s–dimensional Hausdorff measure of $W_p(\tau)$ at the critical exponent $s = n/\tau$ is infinite. For completeness, we mention the following dimension result [**26**]. Let G be a geometrically finite Kleinian group with parabolic elements and let $\mathrm{rk}(p)$ denote the rank of the parabolic fixed point p. Then for $\tau \geq 1$

$$\dim W_p(\tau) \; = \; \min \left\{ \frac{\delta + \mathrm{rk}(p) \, (\tau - 1)}{2\tau - 1} , \; \frac{\delta}{\tau} \right\} .$$

So for groups of the second kind, although the dimension of $W_p(\tau)$ is known its Hausdorff measure at the critical exponent is unknown. As already mentioned above, our general framework fails to shed any light on this, since although for groups of the second kind we are able to establish local m-ubiquity (Proposition 6) the measure m is not of type (M2) or even of type (M2′). Recently, it has been shown that for sets closely related to $W_p(\psi)$ the Hausdorff measure is either zero or infinite [**19**]. However, even for these related sets one is unable to establish the analogue of Theorem 6 for groups of the second kind.

When interpreted on the upper half plane model $\mathbb{H}^2$ of hyperbolic space and applied to the modular group $\mathrm{SL}(2, \mathbb{Z})$, the above theorems imply the classical results associated with our basic example – see §1.1. Next, let G_d denote the Bianchi group of 2x2 matrices of determinate one with entries in the ring of integers $\vartheta = \vartheta(d)$ of the imaginary quadratic field $\mathbb{Q}(\sqrt{-d})$. Here d is a positive integer which is not a perfect square. For a real, positive decreasing function ψ, let $W_\vartheta(\psi)$ denote the set of complex numbers z such that the inequality

$$|z - p/q| \; < \; \psi(|q|)$$

is satisfied for i.m. pairs $p, q \in \vartheta \times \vartheta$ with $\mathrm{ideal}(p,q) = \vartheta$. Following §7 of [**41**], it is easily verified that when interpreted on the upper half space model $\mathbb{H}^3$ of hyperbolic space and applied to the Bianchi group G_d, the above theorems imply the following statements.

THEOREM 7. *Let ψ be a real, positive decreasing function and let m denote 2–dimensional Lebesgue measure. Then*

$$m\left(W_\vartheta(\psi)\right) = \begin{cases} 0 & \text{if } \sum_{r=1}^{\infty} \psi\left(r\right)^2 \ r^3 < \infty\,, \\[2mm] \infty & \text{if } \sum_{r=1}^{\infty} \psi\left(r\right)^2 \ r^3 = \infty\,. \end{cases}$$

THEOREM 8. *Let f be a dimension function such that $r^{-2} f(r) \to \infty$ as $r \to 0$ and $r^{-2} f(r)$ is decreasing. Let ψ be a real, positive decreasing function. Then*

$$\mathcal{H}^f\left(W_\vartheta(\psi)\right) = \begin{cases} 0 & \text{if } \sum_{r=1}^{\infty} f\left(\psi(r)\right) \ r^3 < \infty\,, \\[2mm] \infty & \text{if } \sum_{r=1}^{\infty} f\left(\psi(r)\right) \ r^3 = \infty\,. \end{cases}$$

Theorem 7 is essentially due to Sullivan [**41**]. However, Sullivan assumed a certain regularity condition on ψ. This has been replaced by the more natural condition that ψ is decreasing. Thus, Theorem 7 is the precise analogue of the classical statement of Khintchine. In the case $\psi(r) = r^{-\tau}$ write $W_\vartheta(\tau)$ for $W_\vartheta(\psi)$. Then, Theorem 8 implies the following 'complex' analogue of the Jarnik–Besicovitch theorem.

COROLLARY 7. *For $\tau \geq 2$, $\dim W_\vartheta(\tau) = \frac{4}{\tau}$. Moreover, $\mathcal{H}^{4/\tau}(W_\vartheta(\tau)) = \infty$.*

Convex co-compact groups: These are geometrically finite Kleinian groups without parabolic elements. Thus, the distinguished limit point y is a hyperbolic fixed point. For convex co-compact groups, Patterson measure m is of type (M2) and the situation becomes much more satisfactory.

Let L be the axis of the conjugate pair of hyperbolic fixed points y and y', and let $G_{yy'}$ denote the stabilizer of y (or equivalently y'). Then there is a geometrically motivated set $\mathcal{T}_{yy'}$ of coset representatives of $G_{yy'} \backslash G$; so chosen that if $g \in \mathcal{T}_{yy'}$ then the orbit point $g(0)$ of the origin lies within a bounded hyperbolic distance from the summit s_g of $g(L)$ – the axis of the hyperbolic fixed pair $g(y)$ and $g(y')$. The summit s_g is simply the point on $g(L)$ 'closest' to the origin. For $g \in \mathcal{T}_{yy'}$, let $H_{g(y)}$ be the horoball with base point at $g(y)$ and radius $R_g := 1 - |s_g|$. Then the top of $H_{g(y)}$ lies within a bounded hyperbolic distance of $g(0)$. Furthermore, as a consequence of the definition of $\mathcal{T}_{yy'}$, it follows that $C^{-1} \leq R_g L_g \leq C$ where $C > 1$ is an absolute constant. We are now able to define the subset $\Lambda(\psi)$ of $W_y(\psi)$ in exactly the same way as in the parabolic case with y replacing p and $\mathcal{T}_{yy'}$ replacing $\mathcal{T}_p$.

Essentially the arguments given in [**35**], can easily be modified to obtain the analogue of the local horoball counting result stated above for the parabolic case. We leave the details to the reader. In turn, this enables one to establish Proposition 6 for convex co-compact groups – the statement remains unchanged. Since m is of type (M2) and ρ is u–regular for any $k > 1$, Corollary 2 of Theorem 1 and Theorem 2 yield the divergent parts of the following statements.

THEOREM 9. *Let G be a convex co-compact Kleinian group and y be a hyperbolic fixed point. Let ψ be a real, positive decreasing function. Then*

$$
m(W_y(\psi)) = \begin{cases} 0 & \text{if} \quad \sum_{r=1}^{\infty} \psi\,(r)^{\delta} \ \ r^{\delta-1} < \infty\,, \\[2ex] 1 & \text{if} \quad \sum_{r=1}^{\infty} \psi\,(r)^{\delta} \ \ r^{\delta-1} = \infty\,. \end{cases}
$$

THEOREM 10. *Let G be a convex co-compact Kleinian group and y be a hyperbolic fixed point. Let f be a dimension function such that $r^{-\delta} f(r) \to \infty$ as $r \to 0$ and $r^{-\delta} f(r)$ is decreasing. Let ψ be a real, positive decreasing function. Then*

$$
\mathcal{H}^f (W_y(\psi)) = \begin{cases} 0 & \text{if} \quad \sum_{r=1}^{\infty} f\,(\psi(r)) \ \ r^{\delta-1} < \infty\,, \\[2ex] \infty & \text{if} \quad \sum_{r=1}^{\infty} f\,(\psi(r)) \ \ r^{\delta-1} = \infty\,. \end{cases}
$$

Again, as in the parabolic case, the first of these theorems is not new. However, in previous statements of the theorem a certain regularity condition on the function ψ is assumed. Thus the above Khintchine type theorem is the perfect analogue of the classical statement. Regarding the second theorem, all that was previously known were dimension statements such as $\dim W_y(\psi) = \delta/\tau$ ($\tau \geq 1$) when $\psi(r) = r^{-\tau}$. Clearly, Theorem 10 implies this statement and shows that the s–dimensional Hausdorff measure of $W_y(\psi)$ at the critical exponent $s = \delta/\tau$ is infinite.

12.4. Rational Maps. We consider a special case of the general 'shrinking target' problem introduced in [**24**]. Let T be an expanding rational map (degree ≥ 2) of the Riemann sphere $\overline{\mathbb{C}} = \mathbb{C} \cup \{\infty\}$ and $J(T)$ be its Julia set. For any $z_o \in J(T)$ and ψ an approximating function, consider the set

$$W_{z_0}(\psi) = \{z \in J(T) : T^n(z) \in B\left(z_o, \psi(|(T^n)'(z)|)\right) \text{ for i.m. } n \in \mathbb{N}\}.$$

In view of the bounded distortion property for expanding maps (see Proposition 1, [**24**]), there exists a constant $C = C(T) > 1$ such that the set of points z in $J(T)$ which lie in the ball

$$B\left(y, \frac{\psi(C\,|(T^n)'(y)|)}{C\,|(T^n)'(y)|}\right)$$

for i.m. pairs $(y,n) \in I := \{(y,n) : n \in \mathbb{N} \text{ with } T^n(y) = z_o\}$ is a subset of $W_{z_0}(\psi)$. On the other hand, if we replace C by C^{-1} in the above ball then $W_{z_0}(\psi)$ is a subset of the corresponding set of points. It is now clear that $W_{z_0}(\psi)$ is a lim sup set of the type considered within our framework. The backward orbit of the selected point z_o in $J(T)$ corresponds to the rationals in the classical theory. This set is also the precise analogue of the set of well approximable limit points associated with a Kleinian group.

With reference to our general framework, let $\Omega := J(T)$, $J := I$, $\alpha := (y,n) \in I$, $\beta_\alpha := C\,|(T^n)'(y)|$ and $R_\alpha := y$. Thus, the family $\mathcal{R}$ of resonant sets R_α consists of pre-images of the point z_o. Furthermore, define $\Delta(R_\alpha, \varphi(\beta_\alpha)) := B(y, \varphi(C\,|(T^n)'(y)|)\,)$ and let

$$\Delta_l^u(\varphi, n) := \bigcup_{(y,m) \in J_l^u(n):} B\left(y, \varphi(C\,|(T^m)'(y)|)\right) ,$$

where $J_l^u(n) := \{(y,m) \in I : k^{n-1} < C\,|(T^m)'(y)| \leq k^n\}$ and $\varphi(r) := r^{-1}\,\psi(r)$. Here $k > 1$ is a constant. In view of the discussion above,

$$W_{z_0}(\psi) \supset \Lambda(\varphi) := \limsup_{n \to \infty} \Delta_l^u(\varphi, n) \ .$$

Now let m be Sullivan measure, $\delta = \dim J(T)$ and $\gamma = 0$. Thus m is a non-atomic, δ–conformal probability measure supported on $J(T)$ and since T is expanding it is of type (M2). We have the following statement concerning local ubiquity.

PROPOSITION 7. *The pair $(\mathcal{R}, \beta)$ is a local m–ubiquitous system relative to (ρ, l, u) where for $k \geq k_o$ – a positive constant dependent only on the rational map T,*

$$l_{t+1} = u_t := k^t \ (t \in \mathbb{N}) \quad and \quad \rho : r \to \rho(r) := \text{constant} \times r^{-1} \ .$$

The proposition follows from the following two facts which can be found in [25]. For ease of reference we keep the same notation and numbering of constants as in [25]. For $X \in \mathbb{R}^+$, let $I(X)$ denote the set of pairs $(y, n) \in I$ such that

$$f_n(y) - C_8 \leq X \leq f_{n+1}(y) + C_8 ,$$

where $f_n(y) := \log |(T^n)'(y)|$; i.e. the n-th ergodic sum of $f = \log |T'|$.

- *Constant Multiplicity:* Let $z \in J(T)$. Then there are no more than C_9 pairs $(y, n) \in I(X)$ such that

$$z \in B \left(y, C_{10} \, |(T^n)'(y)|^{-1} \right) .$$

This is the second part of the statement of Lemma 8 in [25].

- *Local Counting Result:* Let B be an arbitrary Euclidean ball centred on a point of $J(T)$. Then there exists a constant $X_o(B)$ such that for $X \geq X_o(B)$

$$\#\{(y, n) \in I(X) : y \in B\} \asymp m(B) \, e^{\delta X} ,$$

where the implied constants are independent of B. This statement is the last line of the proof of Theorem 4 in [25].

To prove the proposition, let $\rho(r) := C_{10} \, r^{-1}$ and B be an arbitrary ball centred on a point of $J(T)$. Then for $k > e^{2 C_8} \, |T'(z_o)|$ and n sufficiently large we have that

$$m\left(B \, \cap \, \bigcup_{(y,m) \in J_l^u(n)} B\left(y, \rho(k^n)\right) \right) \geq m\left(\bigcup_{\substack{(y,m) \in J_l^u(n): \\ y \in \frac{1}{2} B}} B\left(y, \rho(k^n)\right) \right)$$

$$\geq m\left(\bigcup_{\substack{(y,m) \in I(X): \\ y \in \frac{1}{2} B}} B\left(y, \rho(k^n)\right) \right)$$

$$\geq C_9^{-1} \sum_{(y,m) \in I(X): \, y \in \frac{1}{2} B} m\left(B\left(y, \rho(k^n)\right) \right)$$

$$\gg \rho(k^n)^\delta \, \#\{(y, m) \in I(X) : y \in \tfrac{1}{2} B\}$$

$$\gg k^{-n \delta} \, e^{\delta X} \, m(\tfrac{1}{2} B) \gg m(B) ,$$

where $X := n \log k - \log C - C_8$. The dependence of z_o from the size of k can be removed by choosing $k > e^{2\,C_8} \sup\{|T'(z)| : z \in J(T)\}$.

In view of Proposition 7, the fact that the measure m is of type (M2) and that ρ is u-regular for any $k > 1$, Corollary 2 of Theorem 1 and Theorem 2 yield the divergent parts of the following statements.

THEOREM 11. *Let ψ be a real, positive decreasing function. Then*

$$m(W_{z_0}(\psi)) = \begin{cases} 0 & \text{if } \sum_{r=1}^{\infty} \psi(r)^{\delta}\ r^{-1} < \infty\,, \\ 1 & \text{if } \sum_{r=1}^{\infty} \psi(r)^{\delta}\ r^{-1} = \infty\,. \end{cases}$$

THEOREM 12. *Let f be a dimension function such that $r^{-\delta} f(r) \to \infty$ as $r \to 0$ and $r^{-\delta} f(r)$ is decreasing. Let ψ be a real, positive decreasing function. Then*

$$\mathcal{H}^f(W_{z_0}(\psi)) = \begin{cases} 0 & \text{if } \sum_{r=1}^{\infty} f(\psi(r)/r)\ r^{\delta-1} < \infty\,, \\ \infty & \text{if } \sum_{r=1}^{\infty} f(\psi(r)/r)\ r^{\delta-1} = \infty\,. \end{cases}$$

In [24], Hausdorff dimension results for $W_{z_0}(\psi)$ were established. For example, in the case $\psi(r) = r^{-\tau}$ write $W_{z_0}(\tau)$ for the set $W_{z_0}(\psi)$. Then for $\tau \geq 0$, $\dim W_{z_0}(\tau) = \delta/(1+\tau)$. Recently, it has been shown that for $\tau > 0$ the $\delta/(1+\tau)$–dimensional Hausdorff measure of $W_{z_0}(\tau)$ is either zero or infinity [27]. Clearly, Theorem 12 implies the dimension statement and shows that the Hausdorff measure at the critical exponent is actually infinite.

Theorem 11 enables us to deduce the following 'logarithmic law' for orbit approximation. Fix a point z_0 in J and for any other point z in J let $d_n(z)$ denote the distance of $T^n(z)$ from z_0. Then for m-almost all points z in J

$$\limsup_{n \to \infty} \frac{-\log d_n(z)}{\log \log |(T^n)'(z)|} = \frac{1}{\delta}\,.$$

This statement can be viewed as the (expanding) rational map analogue of Sullivan's logarithmic law for geodesics [40, 41].

12.5. Diophantine approximation with restrictions. In a series of papers, G. Harman has considered the problem of obtaining zero-one laws for hybrids of the classical set $W(\psi)$ in which the numerator and denominator of the rational approximates are restricted to various sets of number theoretic interest. We refer the reader to Chapter 6 in [**22**] for a full exposition.

To illustrate the diversity of our main theorems, we consider a specific case in which the number theoretic set is the set of prime numbers $\mathcal{P}$. Thus, let ψ be an approximating function and let

$$W_{\mathcal{P}}(\psi) := \{x \in [0,1] : |x - p/q| < \psi(q) \text{ for i.m. } (p,q) \in \mathcal{P} \times \mathcal{P}\}.$$

With reference to our general framework, let $\Omega := [0,1]$, $J := \{(p,q) \in \mathcal{P} \times \mathcal{P} : p \leq q\}$, $\alpha := (p,q) \in J$, $\beta_\alpha := q$ and $R_\alpha := p/q$. Thus, the family $\mathcal{R}$ of resonant sets R_α consists of rationals p/q with both numerator and denominator prime. Furthermore, define $\Delta(R_\alpha, \psi(\beta_\alpha)) := B(p/q, \psi(q))$ and let

$$\Delta_l^u(\psi, n) := \bigcup_{\substack{(p,q) \in J_l^u(n):}} B(p/q, \psi(q)) = \bigcup_{\substack{q \in \mathcal{P}: \\ 2^{n-1} < q \leq 2^n}} \bigcup_{p \in \mathcal{P}: p \leq q} B(p/q, \psi(q)),$$

where $J_l^u(n) := \{(p,q) \in J : 2^{n-1} < q \leq 2^n\}$. Then

$$W_{\mathcal{P}}(\psi) = \Lambda(\psi) := \limsup_{n \to \infty} \Delta_l^u(\psi, n).$$

Now let m be one–dimensional Lebesgue measure, $\delta = 1$ and $\gamma = 0$. Then a relatively standard analytic argument making use of sieve theory leads to the following local m–ubiquity statement.

PROPOSITION 8. *The pair $(\mathcal{R}, \beta)$ is a local m–ubiquitous system relative to (ρ, l, u), where $l_{t+1} = u_t := 2^t$, $(t \in \mathbb{N})$ and $\rho : r \to \rho(r) := \text{constant} \times (\log r)^2 r^{-2}$.*

The log term in the function ρ is, of course, a consequence of the prime number theorem. The proposition can be deduced from Lemma 6.3 in [**22**]. However, for the details see [**42**]. Thus, in view of the proposition and the fact that the measure m is of type (M2) and that ρ is u-regular, Corollary 2 of Theorem 1 and Theorem 2 yield the divergent parts of the following statements.

THEOREM 13. *Let ψ be a real, positive decreasing function. Then*

$$W_{\mathcal{P}}(\psi) = \begin{cases} 0 & \text{if} \quad \sum_{r=1}^{\infty} \psi(r) \ r \ (\log r)^{-2} < \infty \ , \\ 1 & \text{if} \quad \sum_{r=1}^{\infty} \psi(r) \ r \ (\log r)^{-2} = \infty \ . \end{cases}$$

This zero-one law was first established by Harman. Theorem 1 shows that the statement is in fact a simple consequence of local m-ubiquity.

THEOREM 14. *Let f be a dimension function such that $r^{-1} f(r) \to \infty$ as $r \to 0$ and $r^{-1} f(r)$ is decreasing. Let ψ be a real, positive decreasing function. Then*

$$\mathcal{H}^f\left(W_{\mathcal{P}}(\psi)\right) = \begin{cases} 0 & \text{if} \quad \sum_{r=1}^{\infty} f\left(\psi(r)\right) \ r \ (\log r)^{-2} < \infty \\ \infty & \text{if} \quad \sum_{r=1}^{\infty} f\left(\psi(r)\right) \ r \ (\log r)^{-2} = \infty \ . \end{cases}$$

Consider the case when $\psi(r) = r^{-\tau}$. Then, Theorem 14 implies that $\dim W_{\mathcal{P}}(\psi) = 2/\tau$ $(\tau \geq 2)$ just as in the classical case. However, the $2/\tau$–dimensional Hausdorff measure is zero unlike the classical case in which it is infinite. Thus, restricting to prime numerators and denominators has no effect on the dimension but drastically effects the Hausdorff measure. The above mentioned dimension result is not new and can be found in [**22**, Theorem 10.8].

For analogous results associated with problems in which the numerator and denominator of the rational approximates are restricted to other sets of number theoretic interest see [**42**].

12.6. Diophantine approximation in $\mathbb{Q}_p$. For a prime p, let $| \ |_p$ denote the p–adic metric and let $\mathbb{Q}_p$ denote the p–adic field. Furthermore, let $\mathbb{Z}_p$ denote the ring of p–adic integers and for a vector $\mathbf{x}$ in $\mathbb{Q}_p^m$ let $|\mathbf{x}|_p := \max\{|x_1|_p, \ldots, |x_m|_p\}$ – the p–adic norm on the vector space $\mathbb{Q}_p^m$. We now consider the p–adic analogue of the 'classical' set $W(m, n; \psi)$ – see §12.1. For an approximating function ψ, let

$$W_p(m, n; \psi) := \{X \in \mathbb{Z}_p^{mn} : |\mathbf{q}X + \mathbf{p}|_p < \psi(\max(|\mathbf{p}|, |\mathbf{q}|))$$

$$\text{for i.m. } (\mathbf{p}, \mathbf{q}) \in \mathbb{Z}^n \times \mathbb{Z}^m\},$$

where $|\mathbf{x}| := \max\{|x_1|, \ldots, |x_m|\}$ is the usual supremum norm of a vector $\mathbf{x}$ in $\mathbb{Z}^m$. By definition, $|\mathbf{q}X + \mathbf{p}|_p = \max_{1 \leq j \leq n} |\mathbf{q}.X^{(j)} + p_j|_p$ where $X^{(j)}$ is the j'th column vector of the $m \times n$ matrix $X \in \mathbb{Z}_p^{mn}$.

There are two points worth making when comparing the above set with the 'classical' set $W(m, n; \psi)$. Firstly, the approximating function in the p–adic setup is a function of $\max(|\mathbf{p}|, |\mathbf{q}|)$ rather than simply $|\mathbf{q}|$. This is due to the fact that within the p–adic setup for any $X \in \mathbb{Z}_p^{mn}$ and $\mathbf{q} \in \mathbb{Z}^m$ there exists $\mathbf{p} \in \mathbb{Z}^n$ such that the quantity $|\mathbf{q}X + \mathbf{p}|_p$ can be made arbitrarily small. Thus, the set of $X \in \mathbb{Z}_p^{mn}$ for which $|\mathbf{q}X + \mathbf{p}|_p < \psi(|\mathbf{q}|)$ for i.m. $(\mathbf{p}, \mathbf{q}) \in \mathbb{Z}^n \times \mathbb{Z}^m$ is in fact the whole space $\mathbb{Z}_p^{mn}$ and there is nothing to prove. Secondly, in the p-adic setup the 'normalizing' factor $|\mathbf{q}|$ does not appear on the right hand side of $|\mathbf{q}X + \mathbf{p}|_p$. This is due to the fact that the p–adic metric is an ultra metric. Thus, if $|\mathbf{q}X + \mathbf{p}|_p < \psi(\max(|\mathbf{p}|, |\mathbf{q}|))$ then X lies in the $\psi(\max(|\mathbf{p}|, |\mathbf{q}|))$ neighborhood of $\{X \in \mathbb{Z}_p^{mn} : \mathbf{q}X + \mathbf{p} = 0\}$ – the resonant set associated with the pair $(\mathbf{p}, \mathbf{q})$ (see below).

With reference to our general framework, let $\Omega := \mathbb{Z}_p^{mn}$, $J := \{(\mathbf{p}, \mathbf{q}) \in \mathbb{Z}^n \times \mathbb{Z}^m \backslash \{\mathbf{0}\} : |\mathbf{p}| \leq |\mathbf{q}|\}$, $\alpha := (\mathbf{p}, \mathbf{q}) \in J$, $\beta_\alpha := |\mathbf{q}|$ and $R_\alpha := \{X \in \mathbb{Z}_p^{mn} : \mathbf{q}X + \mathbf{p} = 0\}$. Thus, the family $\mathcal{R}$ of resonant sets R_α consists of $(m-1)n$–dimensional sets. Furthermore, $\Delta(R_\alpha, \psi(\beta_\alpha)) := \{X \in \mathbb{Z}_p^{mn} : \text{dist}(X, R_\alpha) \leq \psi(|\mathbf{q}|)\}$ and let

$$\Delta_l^u(\psi, n) := \bigcup_{2^{n-1} < |\mathbf{q}| \leq 2^n} \bigcup_{|\mathbf{p}| \leq |\mathbf{q}|} \Delta(R_\alpha, \psi(\beta_\alpha)) .$$

The metric d is of course the p–adic norm. Then

$$W(m, n; \psi) \supset \Lambda(\psi) := \limsup_{n \to \infty} \Delta_l^u(\psi, n) .$$

Now let m be the standard Haar measure on $\mathbb{Q}_p^{mn}$, $\delta = mn$ and $\gamma = (m-1)n$. Thus $m(\mathbb{Z}_p^{mn}) = 1$ and $m(B(x, p^{-t})) = (p^{-t})^{mn}$ for $t \in \mathbb{N}$. Then a probabilistic argument involving mean and variance techniques yields the following statement.

PROPOSITION 9. *The pair $(\mathcal{R}, \beta)$ is a local m–ubiquitous system relative to (ρ, l, u) with ρ, l and u as in Proposition 4.*

The ideas necessary for the proof can be found in [**33**, Chp. 4]. Briefly, for $N \in \mathbb{N}$ define the set $\mathcal{F}_N$ to be empty if $|N|_p < 1$ and if $|N|_p = 1$ let $\mathcal{F}_N$ be the set of $(\mathbf{p}, \mathbf{q}) \in \mathbb{Z}^n \times \mathbb{Z}^m$ such that

(1) $|\mathbf{r}| = |\mathbf{q}| = q_1 = N$
(2) p_j is co-prime to q_1 with $0 < p_j < N \omega(N)^{-1/2}$ for $j = 1, \ldots, n$.

The argument on page 86 together with the mean–variance argument beginning on page 94 and Lemma 4.1 on page 70 imply that for n large enough $m\left(\Delta_l^u(\rho,n)\right) \geq \kappa > 0$, where ρ is as in Proposition 9. This proves global m-ubiquity. However, much more is true. A simple applications of Lemma 1.1 on page 14 enables us to conclude that $m\left(\Delta_l^u(\rho,n)\right) \to 1$ as $n \to \infty$. The required local m-ubiquity statement immediately follows – see §2.4.

In view of the proposition and the fact that the measure m is of type (M2) and that ρ is u-regular, Corollary 2 of Theorem 1 and Theorem 2 yield the divergent parts of the following statements. They are the p-adic analogues of the theorems stated in §12.1.

THEOREM 15. *Let ψ be a real, positive decreasing function. Then*

$$m(W_p(m,n;\psi)) = \begin{cases} 0 & \text{if } \sum_{r=1}^{\infty} \psi(r)^n \; r^{m+n-1} < \infty, \\ 1 & \text{if } \sum_{r=1}^{\infty} \psi(r)^n \; r^{m+n-1} = \infty. \end{cases}$$

THEOREM 16. *Let f be a dimension function such that $r^{-mn} f(r) \to \infty$ as $r \to 0$ and $r^{-mn} f(r)$ is decreasing. Furthermore suppose that $r^{-(m-1)n} f(r)$ is increasing. Let ψ be a real, positive decreasing function. Then*

$$\mathcal{H}^f\left(W_p(m,n;\psi)\right) = \begin{cases} 0 & \text{if } \sum_{r=1}^{\infty} f(\psi(r)) \; \psi(r)^{-(m-1)n} \, r^{m+n-1} < \infty, \\ \infty & \text{if } \sum_{r=1}^{\infty} f(\psi(r)) \; \psi(r)^{-(m-1)n} \, r^{m+n-1} = \infty. \end{cases}$$

Theorem 15 is not new and first appeared (in full generality) in [33]; the specific case $m = 1$ is due to Jarník [29]. Theorem 16 is new. Previously [1], it had been shown that $\dim W_p(m,n;\tau) = (m-1)n+(m+n)/\tau$ for $\tau > (m+n)/n$. As usual, $W_p(m,n;\tau)$ corresponds to the set $W_p(m,n;\psi)$ with $\psi : r \to r^{-\tau}$. Obviously, Theorem 16 implies this dimension result and shows that $\mathcal{H}^s(W_p(m,n;\tau)) = \infty$ at the critical exponent $s = \dim W_p(m,n;\tau)$.

12.7. Diophantine approximation on manifolds. Let M denote an m-dimensional submanifold in $\mathbb{R}^n$ with $n \geq 2$. Given an approximating function ψ, the problem is to determine measure theoretic laws for points $\mathbf{x} \in \mathbb{R}^n$ resticted to the manifold M. This restriction means that the points $\mathbf{x} = \{x_1, ..., x_n\}$ of interest consist of dependent variables reflecting the fact that $\mathbf{x} \in M$. The fact that the variables are dependent unlike in the classical setup where $M = \mathbb{R}^n$, introduces major difficulties in attempting to describe the measure theoretic structure of lim sup sets restricted to M. There are two main types of lim sup sets that can be considered.

12.7.1. *Dual/linear approximation on manifolds.* For an approximating function ψ, let

$$W(M; \psi) := \{\mathbf{x} \in M : |\mathbf{q}.\mathbf{x} - p| < |\mathbf{q}|\psi(|\mathbf{q}|) \text{ for i.m. } (p, \mathbf{q}) \in \mathbb{Z} \times \mathbb{Z}^n\},$$

where $\mathbf{x}.\mathbf{y} = x_1 y_1 + \cdots + x_n y_n$ is the normal scalar product of two vectors $\mathbf{x}, \mathbf{y}$ in $\mathbb{R}^n$ and $|\mathbf{x}|$ is the usual supremum norm of the vector $\mathbf{x} \in \mathbb{R}^n$. To make any reasonable progress we impose the condition that the m–dimensional manifold M arises from a non–degenerate map $\mathbf{f} : U \to \mathbb{R}^n$ where U is an open subset of $\mathbb{R}^m$ and $M := \mathbf{f}(U)$. The map $\mathbf{f} : U \to \mathbb{R}^n : \mathbf{u} \mapsto \mathbf{f}(\mathbf{u}) = (f_1(\mathbf{u}), \ldots, f_n(\mathbf{u}))$ is said to be *non–degenerate* at $\mathbf{u} \in U$ if there exists some $l \in \mathbb{N}$ such that $\mathbf{f}$ is l times continuously differentiable on some sufficiently small ball centred at $\mathbf{u}$ and the partial derivatives of $\mathbf{f}$ at $\mathbf{u}$ of orders up to l span $\mathbb{R}^n$. The map $\mathbf{f}$ is *non–degenerate* if it is non–degenerate at almost every (in terms of m–dimensional Lebesgue measure) point in U; in turn the manifold $M = \mathbf{f}(U)$ is also said to be non–degenerate. Geometrically, the non–degeneracy of M at $\mathbf{y}_0 \in M$ means that M deviates from any hyperplane in $\mathbb{R}^n$ that contains $\mathbf{y}_0$.

Consider any ball $B_{\mathbf{u}_0}$ centred at $\mathbf{u}_0$ in U. Then, with reference to our general framework let $\Omega := B_{\mathbf{u}_0}, J := \{(p, \mathbf{q}) \in \mathbb{Z} \times \mathbb{Z}^n \setminus \{\mathbf{0}\} : |p| \le |\mathbf{q}|\}$, $\alpha := (p, \mathbf{q}) \in J$, $\beta_\alpha := |\mathbf{q}|$ and $R_\alpha := \{\mathbf{u} \in B_{\mathbf{u}_0} : \mathbf{q}.f(\mathbf{u}) = p\}$. Thus, the family $\mathcal{R}$ of resonant sets R_α have dimension $m - 1$ and arise from the intersection of M with the $(n-1)$–dimensional hyperplanes given by $\{\mathbf{x} \in \mathbb{R}^n : \mathbf{q}.\mathbf{x} = p$ with $(p, \mathbf{q}) \in J\}$. Furthermore, $\Delta(R_\alpha, \psi(\beta_\alpha)) := \{\mathbf{u} \in B_{\mathbf{u}_0} : \text{dist}(\mathbf{u}, R_\alpha) \le \psi(|\mathbf{q}|)\}$ and

$$\Delta_l^u(\psi, n) := \bigcup_{k^{n-1} < |\mathbf{q}| \le k^n} \bigcup_{|\mathbf{p}| \le |\mathbf{q}|} \Delta(R_\alpha, \psi(\beta_\alpha)) .$$

Here $k > 1$ is a constant. Then

$$W(M; \psi) \supset \Lambda(\psi) := \limsup_{n \to \infty} \Delta_l^u(\psi, n) .$$

Note that

$$\Lambda(\psi) = \{\mathbf{u} \in B_{\mathbf{u}_0} : |\mathbf{q}.f(\mathbf{u}) - p| < |\mathbf{q}|\psi(|\mathbf{q}|) \text{ for i.m. } (p, \mathbf{q}) \in \mathbb{Z} \times \mathbb{Z}^n\}.$$

Now let the measure m be normalised m–dimensional Lebesgue measure on $B_{\mathbf{u}_0}$, $\delta = m$ and $\gamma = m - 1$. Then for almost all $\mathbf{u}_0 \in U$ there exists a corresponding ball $B_{\mathbf{u}_0}$ for which the following local m–ubiquity statement holds.

PROPOSITION 10. *The pair $(\mathcal{R}, \beta)$ is a local m–ubiquitous system relative to (ρ, l, u) where for $k > k_0(B_{\mathbf{u}_0})$ – a positive absolute constant, $l_{t+1} = u_t := k^t$ ($t \in \mathbb{N}$) and $\rho : r \to \rho(r) := \text{constant} \times \rho^{-(n+1)}$.*

This proposition can be deduced from Proposition 3.3 in [6]. In view of the proposition and the fact that the measure m is of type (M2) and that ρ is u-regular, Corollary 2 of Theorem 1 and Theorem 2 yield the divergent parts of the following statements.

THEOREM 17. *Let ψ be a real, positive decreasing function, M be a non-degenerate manifold, m be the induced Lebesgue measure on M. Then*

$$
m(W(M; \psi)) = \begin{cases} 0 & \text{if } \sum_{r=1}^{\infty} \psi(r) \; r^n < \infty \,, \\ m(M) & \text{if } \sum_{r=1}^{\infty} \psi(r) \; r^n = \infty \,. \end{cases}
$$

THEOREM 18. *Let f be a dimension function such that $r^{-m} f(r) \to \infty$ as $r \to 0$ and $r^{-m} f(r)$ is decreasing. Furthermore, suppose that $r^{-(m-1)} f(r)$ is increasing. Let ψ be a real, positive decreasing function. Then*

$$
\mathcal{H}^f(W(M; \psi)) = \infty \qquad \text{if} \qquad \sum_{r=1}^{\infty} f(\psi(r)) \, \psi(r)^{-(m-1)} \, r^n = \infty.
$$

Theorem 17 is not new. The convergence half was independently proved in [5] and [11] and the divergence half in [6]. The convergence part is not at all obvious and requires delicate covering and counting arguments to make use of the 'natural' cover of $W(M; \psi)$. In fact it implies that any non–degenerate manifold is extremal which was a longstanding conjecture of Baker–Sprindzuk. This conjecture was proved independently in [30].

Theorem 18 is new. It shows that any non-degenerate manifold is of Jarník type for divergence in the case of dual approximation where Jarník type is the Hausdorff measure analogue of the notion of Khintchine/Groshev type [10, pg 29]. However, unlike in previous applications, we are currently unable to show that $\mathcal{H}^f(W(M; \psi)) =$

0 when the sum in Theorem 18 converges. Previously [**14**], it had been shown that $\dim W(M; \tau) \geq m - 1 + (n+1)/\tau$ for $\tau > n+1$. Currently there is no general upper bound. As usual $W(M; \tau)$ corresponds to the set $W(M; \psi)$ with $\psi : r \mapsto r^{-\tau}$. Obviously Theorem 18 implies this dimension result and shows that $\mathcal{H}^s(W(M; \tau)) = \infty$ for $s = m - 1 + (n+1)/\tau$ which is almost certainly the critical exponent.

12.7.2. *Simultaneous approximation on manifolds.* For an approximating function ψ, let

$$W(M; \psi) = \{\mathbf{x} \in M : |q\mathbf{x} - \mathbf{p}| < |q|\psi(|q|) \text{ for i.m. } (\mathbf{p}, q) \in \mathbb{Z}^n \times \mathbb{Z}\}.$$

Recall that M is an m–dimensional submanifold embedded in $\mathbb{R}^n$. Even under the restriction that M is non–degenerate, results analogous to those described above for dual approximation currently seem out of reach. However in [**8**], we have recently made advances in the case that M is a C^3 planar curve. Moreover, for particular planar curves such as the unit circle and the parabola we are able to establish reasonably complete measure theoretic laws. To this end, let $M = S^1 := \{(x_1, x_2) \in \mathbb{R}^2 : x_1^2 + x_2^2 = 1\}$ denote the unit circle. Thus, $W(S^1; \psi)$ consists of points $(x_1, x_2) \in S^1$ for which there exist i.m. rational pairs $(\frac{p_1}{q}, \frac{p_2}{q})$ such that the following pair of inequalities are simultaneously satisfied:

(62) $|x_i - p_i/q| < \psi(|q|)$ $(i = 1, 2)$.

Throughout the following discussion, assume that $r^2\psi(r) \to 0$ as $r \to \infty$. Also, without loss of generality assume that $q \in \mathbb{N}$. The following fact shows that under the above assumption on ψ, there is a one to one correspondence between the rational approximates $(p_1/q, p_2/q)$ satisfying (62) and the Pythagorean triples $s^2 + t^2 = q^2$.

- For q large, any rational pair $(p_1/q, p_2/q)$ satisfying (62) lies on S^1.

This is trivial, since $(1 - \psi(q))^2 \leq (p_1/q)^2 + (p_2/q)^2 < (1 + 3\psi(q))^2$ and so for q sufficiently large $|p_1^2 + p_2^2 - q^2| < 1$. Now notice that the left hand side is an integer.

Now with reference to our general framework, let $\Omega := S^1, J := \{(\mathbf{p}, q) \in \mathbb{Z}^2 \times \mathbb{N} : \mathbf{p}/q \in S^1\}$, $\alpha := (\mathbf{p}, q) \in J$, $\beta_\alpha := q$ and $R_\alpha := \mathbf{p}/q$. Thus, the family $\mathcal{R}$ of resonant sets R_α consists of rational pairs $(p_1/q, p_2/q)$ lying on S^1. Furthermore, define $\Delta(R_\alpha, \psi(\beta_\alpha)) := B(\mathbf{p}/q, \psi(q))$ and let

$$\Delta_l^u(\psi, n) := \bigcup_{2^{n-1} < q \leq 2^n} \bigcup_{\substack{\mathbf{p} \in \mathbb{Z}^2: \\ p_1^2 + p_2^2 = q^2}} B(\mathbf{p}/q, \psi(q)) \,,$$

Then

$$W(S^1; \psi) \supset \Lambda(\psi) := \limsup_{n \to \infty} \Delta_l^u(\psi, n) \,.$$

Now let m be normalized Lebesgue measure on S^1, $\delta = 1$ and $\gamma = 0$. Then we have the following local m–ubiquity statement.

PROPOSITION 11. *The pair $(\mathcal{R}, \beta)$ is a local m–ubiquitous system relative to (ρ, l, u), where $l_{t+1} = u_t := 2^t$ $(t \in \mathbb{N})$ and $\rho : r \to \rho(r) := \text{constant} \times r^{-1}$.*

Although this statement can be found explicitly in [15], we shall give an alternative proof which is shorter and probably more adaptable to analogous problems. For a point a on S^1, let A be the arc with centre (mid-point) a and radius $r(A)$. Clearly $r(A) \asymp m(A)$. For $N \in \mathbb{N}$, let $\mathcal{C}(N, A) := \{(\mathbf{p}, q) \in J : \mathbf{p}/q \in A \text{ with } N < q \leq 2N\}$. The proposition is a simple consequence of the following two facts.

• For $N \geq N_o(A)$, $\#\mathcal{C}(N, A) \asymp r(A) N \asymp m(A) N$.

This follows from standard results on the distribution of Pythagorean triples.

• For $N \geq N_o$, if $r(A) < 2^{-4/3} N^{-1}$ then $\#\mathcal{C}(N, A) \leq 2$.

This is the key result. Let b and b' be the end points of A. Then for N large enough, the triangle T subtended by the three points a, b, b' has area less than $r(A)^3$. Now suppose there are three rational points $\mathbf{s}/q, \mathbf{t}/q', \mathbf{u}/q'' \in \mathcal{C}(N, A)$ and let $\triangle$ be the triangle subtended by them. Clearly, $\text{area}(\triangle) \leq \text{area}(T)$. Thus

$$2\, r(A)^3 \geq 2\, \text{area}(\triangle) = \begin{vmatrix} 1 & s/q & t/q \\ 1 & s'/q' & t'/q' \\ 1 & s''/q'' & t''/q'' \end{vmatrix} \geq \frac{1}{q\,q'q''} \geq \frac{1}{(2N)^3} \,.$$

Hence if $r(A) < 2^{-4/3} N^{-1}$, the triangle $\triangle$ cannot exist so the three rational points must lie on a straight line. However this is impossible since they lie on S^1.

Thus, in view of the proposition and the fact that the measure m is of type (M2) and that ρ is u-regular, Theorem 2 yields the divergent parts of the following statement.

THEOREM 19. *Let f be a dimension function such that $r^{-1} f(r) \to \infty$ as $r \to 0$ and $r^{-1} f(r)$ is decreasing. Let ψ be a real, positive decreasing function such that $r^2 \psi(r) \to 0$ as $r \to \infty$. Then*

$$\mathcal{H}^f\left(W(S^1; \psi)\right) = \begin{cases} 0 & \text{if } \sum_{r=1}^{\infty} f(\psi(r)) < \infty \,, \\ \infty & \text{if } \sum_{r=1}^{\infty} f(\psi(r)) = \infty \,. \end{cases}$$

Let us consider the case when $\psi(r) = r^{-\tau}$ and write $W(S^1; \tau)$ for $W(S^1; \psi)$. In [15], it was shown that $\dim W(S^1 \tau) = 1/\tau$ when $\tau > 2$. Clearly, Theorem 19 implies this dimension result and shows that $\mathcal{H}^{1/\tau}(W(S^1; \tau)) = \infty$. Note that the condition that $\tau > 2$, ensures that $r^2 \psi(r) \to 0$ as $r \to \infty$ and so that the rational points of interest are forced to lie on the circle. This is not the case when $\tau \leq 2$ and even the problem of determining the dimension of $W(S^1; \tau)$ becomes highly non-trivial. In view of Dirchlet's theorem on simultaneous Diophantine approximation one knows that $W(S^1; \tau) = S^1$ for $\tau \leq 3/2$. On the other hand, it is not difficult to show that $m(W(S^1; \tau)) = 0$ for $\tau > 3/2$ and moreover that $\dim W(S^1; \tau) \leq (3 - \tau)/\tau$ when $3/2 \leq \tau \leq 2$.

In a forthcoming paper [8], general measure theoretic laws for non-degenerate C^3 planar curves are established. A simple consequence of these results is that:

$$\dim W(S^1; \tau) = (3 - \tau)/\tau \qquad (3/2 \leq \tau \leq 2) \,.$$

Another consequence of the results in [8] is the following law with respect to the measure m on S^1.

THEOREM. *Let ψ be a real, positive decreasing function. Then*

$$m\left(W(S^1; \psi)\right) = \begin{cases} 0 & \text{if } \sum_{r=1}^{\infty} (r\,\psi(r))^2 < \infty \,, \\ 1 & \text{if } \sum_{r=1}^{\infty} (r\,\psi(r))^2 = \infty \,. \end{cases}$$

12.8. Sets of exact order. With reference to our general framework, given two approximating functions φ and ψ with φ in some sense 'smaller' than ψ, consider the set $E(\psi, \varphi)$ of points x in Ω for which

$$x \in \Delta(R_\alpha, \psi(\beta_\alpha)) \quad \text{for i.m. } \alpha \in J;$$

and that

$$x \notin \Delta(R_\alpha, \varphi(\beta_\alpha)) \quad \text{for all but finitely many } \alpha \in J \ .$$

In short, $E(\psi, \varphi) := \Lambda(\psi) \setminus \Lambda(\varphi)$. Thus the approximation properties of points x in $E(\psi, \varphi)$ are 'sandwiched' between the functions φ and ψ. In [7], under the classical linear forms setup (cf. §12.1), we have shown that the measure theoretic laws for $W(m, n; \psi)$ with respect to the measures m (Theorem (Khintchine–Groshev)) and $\mathcal{H}^f$ (Theorem DV) give rise to precise metric results for the corresponding 'exact order' sets $E(m, n; \psi, \varphi)$. In short, the key idea is to construct an appropriate dimension function f for which $\mathcal{H}^f(W(m, n; \psi)) = \infty$ and $\mathcal{H}^f(W(m, n; \varphi)) = 0$ and so $\mathcal{H}^f(E(m, n; \psi, \varphi)) = \infty$.

Regarding our general framework and the exact order sets $E(\psi, \varphi)$, the arguments used in [7] can be carried over to obtain analogous statements of the theorems in [7] provided analogues of both the Khintchine–Groshev Theorem and Theorem DV hold. The point is that both the divergence and convergence halves of these theorems are required. In any case, this is the case for all the applications considered in this section except the previous one and the Kleinian groups application when the group is of the second kind with parabolic elements. Thus, for each of the other applications it is possible to describe the measure theoretic structure of the associated exact order sets $E(\psi, \varphi)$. We shall leave the details to the energetic reader.

Acknowledgements. SV would like to thank his new friends, Ayesha and Iona for their permanent smiles and wonderfully positive 'outlook' – long may it last !! Finally and most importantly of all he would like to thank Bridget Bennett for ... just about everything.

We would like to thank the reviewer for her/his comments which have made the early part of the paper more accessible and for the reference to Knopp [31]. We would also like to thank the editor for his professionalism and 'speed of response' during the various stages of the reviewing process.

Bibliography

[1] A. Abercrombie : The Hausdorff dimension of some exceptional sets of p–adic integer matrices. *J. Number Theory* (1995) **53**, 311–341.

[2] A. Baker and W. M. Schmidt : Diophantine approximation and Hausdorff dimension. *Proc. Lond. Math. Soc.*, 21 (1970) 1–11.

[3] V. V. Beresnevich : On approximation of real numbers by real algebraic numbers. *Acta Arith.* 90 (1999) 97-112.

[4] V.V.Beresnevich : Application of the concept of regular systems in the Metric theory of numbers. *Vestsi Nats. Acad. Navuk Belarusi. Ser. Fiz.-Mat. Navuk*, in Russian, 1 (2000) 35–39.

[5] V. V. Beresnevich : A Groshev type theorem for convergence on manifolds. *Acta Math. Hungarica* **94** (2002) 99–130.

[6] V. V. Beresnevich, V. I. Bernik, D. Y. Kleinbock and G. A. Margulis : Metric Diophantine approximation: the Khintchine–Groshev Theorem for non–degenerate manifolds. *Moscow Math. Jou.* **2** (2002) 203–225.

[7] V. V. Beresnevich, H. Dickinson and S. L. Velani : Sets of exact 'logarithmic' order in the theory of Diophantine approxiamtion. *Math. Ann.* 321 (2001) 253–273.

[8] V. V. Beresnevich, H. Dickinson and S. L. Velani : Diophantine approxiamtion on planer curves and the distribution of rational points. With an Appendix by R. C. Vaughan : Sums of two squares near perfect squares. Pre-print. arXiv:mathNT/0401148 (2004) 1–39.

[9] V. V. Beresnevich and S. L. Velani : A Mass Transference Principle and the Duffin–Schaeffer conjecture for Hausdorff measures. Pre-print. arXiv:mathNT/0412141 (2004) 1–22.

[10] V. I. Bernik and M. M. Dodson : *Metric Diophantine approximation on manifolds.* Cambridge Tracts in Mathematics 137, C.U.P., (1999).

[11] V. I. Bernik, D. Kleinbock and G. A. Margulis : Khintchine–type theorems on manifolds: the convergence case for standard and multiplicative versions. *IMRN* **9** (2001) 453–486.

[12] Y. Bugeaud : Approximation par des nombres algébriques de degré borné at dimension de Hausdorff. *J. Number Theory* 96 (2002) 174–200.

[13] Y. Bugeaud : An inhomogeneous Jarník theorem. Pre-print (2003) 1–21.

[14] H. Dickinson and M. M. Dodson : Extremal manifolds and Hausdorff dimension. *Duke Math. J.*, 101 (2000) 271–281.

[15] H. Dickinson and M. M. Dodson : Diophantine approximation and Hausdorff dimension on the circle. *Math. proc. Camb. Phil.*, 130 (2001) 515–522.

[16] H. Dickinson and S. L. Velani : Hausdorff measure and linear forms. *J. reine angew. Math.*, 490 (1997) 1-36.

[17] M.M. Dodson, M.V. Melián, D. Pestana and S. L. Velani : Patterson measure and Ubiquity. *Ann. Acad. Sci. Fenn.*, 20:1 (1995) 37–60.

[18] M. M. Dodson, B. P. Rynne and J. A. G. Vicker : Diophantine approximation and a lower bound for Hausdorff dimension. *Mathematika*, 37 (1990) 59–73.

[19] M. M. Dodson and S. L. Velani : in preperation.

[20] K. Falconer : *Fractal Geometry: Mathematical Foundations and Applications.* John Wiley & Sons, (1990).

[21] H. Federer : *Geometric Measure Theory.* Sringer-Verlag, (1969).

[22] G. Harman : *Metric Number Theory.* LMS Monographs 18, Clarendon Press, Oxford, (1998).

[23] J. Heinonen : *Lectures on analysis on metric spaces.* Universitext, Springer – Verlag, (2001).

[24] R. Hill and S. L. Velani : Ergodic theory of shrinking targets. *Inventiones mathematicae*, 119 (1995) 175-198.

[25] R. Hill and S.L. Velani : Metric Diophantine approximation in Julia sets of expanding rational maps. Inst. Hautes Etudes Sci. Publ. Math., 85 (1997) 193-216.

[26] R. Hill and S. L. Velani : The Jarník -Besicovitch theorem for geometrically finite Kleinian groups, *Proc. Lond. Math. Soc.*, (3) 77 (1998) 524-550.

[27] R. Hill and S. L. Velani : A Zero–Infinity Law for well approximable points in Julia sets, *Ergodic Theory & Dyn. Syst.*, 22 (2002) 1773–1782.

[28] I. Jarník : Zur metrischen Theorie der diophantischen Appoximationen. *Proc. Mat. Fyz.*, 36 (1928) 91–106.

[29] I. Jarník : Sur les approximations diophantiennes des nombres p–adiques. *revista Ci Lima.* **47**, 489–505.

[30] D. Y. Kleinbock and G. A. Margulis : Flows on homeogeneous spaces and Diophantine approximation on manifolds. *Ann. Math.* 148 (1998) 339–360.

[31] K. Knopp : Mengentheoretische Behandlung einiger Probleme der diophantischen Approximationen und der transfiniten Wahrscheinlichkeiten. *Math. Ann.* 95 (1926) 409–426.

[32] J. Levesley : A general inhomogeneous Jarnk-Besicovitch theorem. *J. Number Theory* 71 (1998) 65–80.

[33] E. Lutz : *Sur les approximations Diophantiennes linéaire p–adiques.* Pulications de l'Institut de Mathématique de l'Université de Strasbourg XII, Herman et Cie, Paris, (1955).

[34] P. Mattila : *Geometry of Sets and Measures in Euclidean Spaces.* Cambridge studies in advanced mathematics 44, C.U.P., (1995).

[35] M.V Melián and S. L. Velani : Geodesic excursions into cusps in infinite volume hyperbolic manifolds, *Mathematica Gottingensis*, 45 (1993), 1-22.

[36] S. J. Patterson : Diophantine approximation in Fuchsian groups. *Phil. Trans. Soc. London*, 282 (1976),527-563.

[37] A.D. Pollington and R.C. Vaughan : The k-dimesional Duffin and Schaeffer conjecture. *Mathematika*, 37, (1990) 190-200.

[38] W.M. Schmidt : *Diophantine approximation.* Lecture notes in Math. 785, Springer – Verlag, (1975).

[39] V.G. Sprindžuk : *Metric theory of Diophantine approximation* (translated by R. A. Silverman). V. H. Winston & Sons, Washington D.C. (1979).

[40] B. Stratmann and S. L. Velani : The Patterson measure for geometrically finite groups with parabolic elements, new and old, *Proc. Lond. Math. Soc.*, (3) 71 (1995) 197-220.

[41] D. Sullivan : Disjoint spheres, approximation by imaginary quadratic numbers and the logarithm law for geodesics. *Acta Math*, 149 (1982) 215-37.

[42] R. Thorn : Metric Number Theory: the good and the bad. Ph.D Thesis (2004): Queen Mary, University of London.

[43] S. L. Velani : Geometrically finite groups, Khintchine-type theorems and Hausdorff dimension. *Math. Proc. Cam. Phil. Soc.*, 120 (1996), 647-662.

Editorial Information

To be published in the *Memoirs*, a paper must be correct, new, nontrivial, and significant. Further, it must be well written and of interest to a substantial number of mathematicians. Piecemeal results, such as an inconclusive step toward an unproved major theorem or a minor variation on a known result, are in general not acceptable for publication. Papers appearing in *Memoirs* are generally at least 80 and not more than 200 published pages in length. Papers less than 80 or more than 200 published pages require the approval of the Managing Editor of the Transactions/Memoirs Editorial Board.

As of September 30, 2005, the backlog for this journal was approximately 14 volumes. This estimate is the result of dividing the number of manuscripts for this journal in the Providence office that have not yet gone to the printer on the above date by the average number of monographs per volume over the previous twelve months, reduced by the number of volumes published in four months (the time necessary for preparing a volume for the printer). (There are 6 volumes per year, each containing at least 4 numbers.)

A Consent to Publish and Copyright Agreement is required before a paper will be published in the *Memoirs*. After a paper is accepted for publication, the Providence office will send a Consent to Publish and Copyright Agreement to all authors of the paper. By submitting a paper to the *Memoirs*, authors certify that the results have not been submitted to nor are they under consideration for publication by another journal, conference proceedings, or similar publication.

Information for Authors

Memoirs are printed from camera copy fully prepared by the author. This means that the finished book will look exactly like the copy submitted.

The paper must contain a *descriptive title* and an *abstract* that summarizes the article in language suitable for workers in the general field (algebra, analysis, etc.). The *descriptive title* should be short, but informative; useless or vague phrases such as "some remarks about" or "concerning" should be avoided. The *abstract* should be at least one complete sentence, and at most 300 words. Included with the footnotes to the paper should be the 2000 *Mathematics Subject Classification* representing the primary and secondary subjects of the article. The classifications are accessible from `www.ams.org/msc/`. The list of classifications is also available in print starting with the 1999 annual index of *Mathematical Reviews*. The Mathematics Subject Classification footnote may be followed by a list of *key words and phrases* describing the subject matter of the article and taken from it. Journal abbreviations used in bibliographies are listed in the latest *Mathematical Reviews* annual index. The series abbreviations are also accessible from `www.ams.org/publications/`. To help in preparing and verifying references, the AMS offers MR Lookup, a Reference Tool for Linking, at `www.ams.org/mrlookup/`. When the manuscript is submitted, authors should supply the editor with electronic addresses if available. These will be printed after the postal address at the end of the article.

Electronically prepared manuscripts. The AMS encourages electronically prepared manuscripts, with a strong preference for $\mathcal{A}_{\mathcal{M}}\mathcal{S}$-LaTeX. To this end, the Society has prepared $\mathcal{A}_{\mathcal{M}}\mathcal{S}$-LaTeX author packages for each AMS publication. Author packages include instructions for preparing electronic manuscripts, the *AMS Author Handbook*, samples, and a style file that generates the particular design specifications of that publication series. Though $\mathcal{A}_{\mathcal{M}}\mathcal{S}$-LaTeX is the highly preferred format of TeX, author packages are also available in $\mathcal{A}_{\mathcal{M}}\mathcal{S}$-TeX.

Authors may retrieve an author package from e-MATH starting from `www.ams.org/tex/` or via FTP to `ftp.ams.org` (login as `anonymous`, enter username as password, and type `cd pub/author-info`). The *AMS Author Handbook* and the *Instruction Manual* are available in PDF format following the author packages link from `www.ams.org/tex/`. The author package can be obtained free of charge by sending email to pub@ams.org (Internet) or from the Publication Division, American Mathematical Society, 201 Charles St., Providence, RI 02904, USA. When requesting an author package, please specify $\mathcal{AMS}$-LaTeX or $\mathcal{AMS}$-TeX, Macintosh or IBM (3.5) format, and the publication in which your paper will appear. Please be sure to include your complete mailing address.

Sending electronic files. After acceptance, the source file(s) should be sent to the Providence office (this includes any TeX source file, any graphics files, and the DVI or PostScript file).

Before sending the source file, be sure you have proofread your paper carefully. The files you send must be the EXACT files used to generate the proof copy that was accepted for publication. For all publications, authors are required to send a printed copy of their paper, which exactly matches the copy approved for publication, along with any graphics that will appear in the paper.

TeX files may be submitted by email, FTP, or on diskette. The DVI file(s) and PostScript files should be submitted only by FTP or on diskette unless they are encoded properly to submit through email. (DVI files are binary and PostScript files tend to be very large.)

Electronically prepared manuscripts can be sent via email to pub-submit@ams.org (Internet). The subject line of the message should include the publication code to identify it as a Memoir. TeX source files, DVI files, and PostScript files can be transferred over the Internet by FTP to the Internet node e-math.ams.org (130.44.1.100).

Electronic graphics. Comprehensive instructions on preparing graphics are available at www.ams.org/jourhtml/graphics.html. A few of the major requirements are given here.

Submit files for graphics as EPS (Encapsulated PostScript) files. This includes graphics originated via a graphics application as well as scanned photographs or other computer-generated images. If this is not possible, TIFF files are acceptable as long as they can be opened in Adobe Photoshop or Illustrator. No matter what method was used to produce the graphic, it is necessary to provide a paper copy to the AMS.

Authors using graphics packages for the creation of electronic art should also avoid the use of any lines thinner than 0.5 points in width. Many graphics packages allow the user to specify a "hairline" for a very thin line. Hairlines often look acceptable when proofed on a typical laser printer. However, when produced on a high-resolution laser imagesetter, hairlines become nearly invisible and will be lost entirely in the final printing process.

Screens should be set to values between 15% and 85%. Screens which fall outside of this range are too light or too dark to print correctly. Variations of screens within a graphic should be no less than 10%.

Inquiries. Any inquiries concerning a paper that has been accepted for publication should be sent directly to the Electronic Prepress Department, American Mathematical Society, 201 Charles St., Providence, RI 02904, USA.

Titles in This Series

For a complete list of titles in this series, visit the
AMS Bookstore at **www.ams.org/bookstore/**.